# THE DYNAMIC HUMAN

## Maciej Henneberg
## &
## Arthur Saniotis

*School of Medicine, The University of Adelaide*
*Australia*
*&*
*Institute for Evolutionary Medicine, University of Zurich*
*Switzerland*

# THE DYNAMIC HUMAN

Authors: Maciej Henneberg & Arthur Saniotis

ISBN (eBook): 978-1-68108-235-6

ISBN (Print): 978-1-68108-236-3

First published in 2016.

## Acknowledgements:

Various parts of this book were written while Maciej Henneberg was working at the Universities of Cape Town and the Witwatersrand, Johannesburg in South Africa and achieved their final form when Arthur Saniotis joined Maciej at the University of Adelaide and they both became associates of the Institute of Evolutionary Medicine at the University of Zurich, Switzerland. The first chapters also benefitted from discussions with John (Jack) Schofield, a journalist, co-author of Maciej's other books. Renata J Henneberg, an experienced human biology author herself, read the final text and provided editorial comments.

advertisements or ideas contained in the Work.

## *Limitation of Liability:*

In no event will Bentham Science Publishers, its staff, editors and/or authors, be liable for any damages, including, without limitation, special, incidental and/or consequential damages and/or damages for lost data and/or profits arising out of (whether directly or indirectly) the use or inability to use the Work. The entire liability of Bentham Science Publishers shall be limited to the amount actually paid by you for the Work.

## General:

1. Any dispute or claim arising out of or in connection with this License Agreement or the Work (including non-contractual disputes or claims) will be governed by and construed in accordance with the laws of the U.A.E. as applied in the Emirate of Dubai. Each party agrees that the courts of the Emirate of Dubai shall have exclusive jurisdiction to settle any dispute or claim arising out of or in connection with this License Agreement or the Work (including non-contractual disputes or claims).
2. Your rights under this License Agreement will automatically terminate without notice and without the need for a court order if at any point you breach any terms of this License Agreement. In no event will any delay or failure by Bentham Science Publishers in enforcing your compliance with this License Agreement constitute a waiver of any of its rights.
3. You acknowledge that you have read this License Agreement, and agree to be bound by its terms and conditions. To the extent that any other terms and conditions presented on any website of Bentham Science Publishers conflict with, or are inconsistent with, the terms and conditions set out in this License Agreement, you acknowledge that the terms and conditions set out in this License Agreement shall prevail.

**Bentham Science Publishers Ltd.**
Executive Suite Y - 2
PO Box 7917, Saif Zone
Sharjah, U.A.E.
Email: subscriptions@benthamscience.org

**BENTHAM
SCIENCE**

# CONTENTS

# FOREWORD

Books sometimes are said to "meet a need" or "fill a niche" or call forth a suggestion that the work can address some modest subject area for a limited reading audience that so far has not been served by related works in a circumscribed field. Far more rare are the occasional books that create a space that is all their own and mark out a territory so large that it will be explored fully and tenanted to full extent only later by others. *The Dynamic Human* does just that: it creates a new domain. In doing so it brings to mind Julian Huxley's *Evolution: the modern synthesis* (Allen & Unwin, London, 1942), which did indeed join together much particularistic biological knowledge that had been amassed before, and organized it into a framework that would inspire and guide several generations of scientists through places that they had not even imagined before. The effect is to create broad avenues of influence among fields that previously had been linked by wandering footpaths or had existed entirely in isolation.

The territory mapped out here begins by establishing a solid core of knowledge about the place of humans in nature, and their ancestry ranging from the earliest upright and bipedal humans such as *Orrorin tugenensis* at about six million years ago through our immediate Neandertal predecessors and thereby to ourselves. The treatment of this material is dynamic indeed; the fossils that comprise tangible anchor points are not dogmatically arrayed into socially constructed "just so" stories describing the supposed "bushiness" of our ancestral record. Instead, the complexity – and, necessarily, uncertainty – of interpretation is signaled by noting that there are nearly two dozen different definitions of the species category, considered by many to be a routinely reliable building block, which it is not. Interpretation of ancestral relationships is informed further by introducing principles of microevolution known from living populations. These are offered as the basis for understanding more reliably the past, during which the same mechanisms must have operated to bring about major transformations over long spans of time.

Evolution of body leads enchantingly here into exploration of the mind, biosphere melding into noosphere engendered by the internet and all of its quantum informational permutations. In this insightful projection of our human future, multiple modes of thought receive their due. Allowance is made for the intuitive approaches of shamans along with the systematically logical approaches more familiar to scientists. Particularly valuable are the concluding sections exploring the brain and cognitive enhancement technologies. Humans have a somatic past that has been shaped by millions of years of organic evolution. From it our minds have emerged to operate *via* thought processes both conscious and unconscious, imaginative as well as analytical. Now brain-machine interfaces are upon us, with at least 59,000 people already recipients of neurological enhancement devices. In the future, more minds will exhibit mental processes benefiting from these and other cognitive enhancement technologies. More numerous by orders of magnitude are our conspecifics whose minds are altered by a plethora of pharmaceuticals. In this "brave new world", the challenges we face will include the need to deal with the increasingly blurred boundaries between reality and fiction. *The Dynamic Human* will help many of us make sense of our chaotically exciting world.

**Robert B. Eckhardt**
Pennsylvannia State University
State College
PA 16801
United States

# PREFACE

Not everyone accepts organic evolution. Those who accept it often see it as a list of events that occurred in the past and produced fossil entities. In relation to our own origins, if we accept evolution at all, we see it as a creative force that must have produced "the first human" at a particular time and in a particular place. Lots of research effort has been expended to pinpoint the precise date and location of this event. Once we have learned when and where our species emerged we think it has been complete and the rest is just learning how it spread around the world and shaped its history. We are studying its characteristics in order to better understand how to save and prolong its life. There is still a tendency to view human beings as a static category that will continue as such into the future unless some catastrophe causes its extinction. The same train of thought makes us believe that we are all copies of the same template that can be understood by studying what is typical.

There is an alternate approach that views the world as a continuously changing complex system comprising variable units that do not conform to any stable plan. We humans are a part of this interminably changing system. We did not appear suddenly and we are not resistant to change.

Like other organisms, human animals continue to evolve generation by generation. As noted by many thinkers, evolution is a non-linear process that is notoriously indeterminate. No two organisms, including humans, are exactly alike and each generation differs, albeit sometimes imperceptibly, from the previous and the next generation.

Human mind has a biological substrate. It is not just the very physical structure of the brain with its maze of interconnected nerve cells, but also the chemical regulation of the entire body that changes the way nerve cells communicate. Therefore, the entire body informs the mind. Our bodies are suffused in the rhythms of nature. The body is a plenum of kaleidoscopic interactions. Through this, individual minds communicate with nature and with each other. This is an interaction borne out of millions of years of trial and error embedded in nature. The human mind is not a logical machine, it is a product of organic interactions.

Our present-day existence is but a short stop in the journey of our ancestors from the past into the future. While our technologies may continue to inform the journey of human bodies and minds, they are incapable of arresting it.

## CONFLICT OF INTEREST

The entire work on the manuscript of this book has not been supported by any grants or other financial contributions and thus there are no conflicts of interest to declare, while the contents are a product of free intellectual endeavour of the authors.

## ACKNOWLEDGEMENTS

Various parts of this book were written while Maciej Henneberg was working at the Universities of Cape Town and the Witwatersrand, Johannesburg in South Africa and achieved their final form when Arthur Saniotis joined Maciej at the University of Adelaide and they both became associates of the Institute of Evolutionary Medicine at the University of Zurich, Switzerland. The first chapters also benefitted from discussions with John (Jack) Schofield, a journalist, co-author of Maciej's other books. Renata J Henneberg, an experienced human biology author herself, read the final text and provided editorial comments.

We thank Tavik Morgenstern for photography for the book cover, Urwah Nawaz for modeling for this photography and Meghan McKinnon for providing child's face reconstruction.

**Maciej Henneberg and Arthur Saniotis**
Biological Anthropology and Comparative Anatomy Unit
The University of Adelaide, Australia and
The Institute of Evolutionary Medicine
University of Zurich, Switzerland

# THE DYNAMIC HUMAN

# CHAPTER 1

# Introduction: Our Place in Nature

**Abstract:** In this book, the authors offer alternative views based on the relativity of time and the cyclic nature of processes affecting our world. We contend that human evolution has been a gradual, though running at various speeds, process during which a complex system of feedback loops has led to enhancement of some human characteristics and the loss of others. Overall, these feedbacks have been of a self-amplifying nature which allows exponential change to occur. The same feedbacks can, however, be reversed by minor alterations in the rates of various natural and cultural cyclic processes that can produce stable states or even declines, depending on how their delicate balances are influenced.

We do not presume to predict events. Our aim is to offer a means of understanding the network of relationships connecting our biological make-up, our technologies, our social structures and the nature of the universe - a framework for better understanding of the current human condition and for construction of possible future scenarios.

**Keywords:** Climate change, Positive feedbacks, Universe.

The most common current tenet among the educated public is the "Big Bang" origin of the universe and the contention that the cosmos has since been expanding, it will slow and reverse in the future, and eventually contract to a single point – a singularity marking the end of everything. Recently, it is being postulated that our universe is but one of many universes filling a bigger entity – the multiverse. What the multiverse fills up together with other multiverses has not been explained yet.

At some point in the evolution of our universe, the story goes, humans appeared either as a natural consequence of organic evolution or as a result of special creation.

Maciej Henneberg & Arthur Saniotis

From that disputed singular point onwards, human history has progressed from hunting-gathering to simple food production, to organised states, well-structured empires and industrialisation - all of which have led to the gradual conquering of the natural world and the beginning of human exploration of space, in the meantime encountering threats of climate change, economic crisis and so forth.

With such an implied simplistic and linear view of history these threats can end in only one of three ways: in complete catastrophe, in collective human wisdom initiating a corrective response which perpetuates our existence and progress - or we all go to Heaven or Hell outside of our universe.

In this book, the authors offer alternative views based on the relativity of time and the cyclic nature of processes affecting our world. We contend that human evolution has been a gradual, though running at various speeds, process during which a complex system of feedback loops has led to enhancement of some human characteristics and the loss of others. Overall, these feedbacks have been of a self-amplifying nature which allows exponential change to occur. The same feedbacks can, however, be reversed by minor alterations in the rates of various natural and cultural cyclic processes that can produce stable states or even declines, depending on how their delicate balances are influenced.

We do not presume to predict events. Our aim is to offer a means of understanding the network of relationships connecting our biological make-up, our technologies, our social structures and the nature of the universe - a framework for better understanding of the current human condition and for construction of possible future scenarios.

We will not attempt to disprove any theology or dismiss any possibility simply because it either conforms to the accepted view or offends somebody, and we intend to postulate only within the bounds of scientific plausibility. We leave the study of the existence and nature of a universal divinity to those more qualified without adopting a stance one way or another. Yet, as you will see, we conclude that almost anything is possible when speculation is involved.

We will endeavour to sketch development of humanity and offer keys to understanding what it means to be human and continue to be human in the future.

# CHAPTER 2

# Origins: A Short History of a Long Journey

**Abstract:** In this chapter, the briefest possible description of changes that are crucial for differentiation of humans from their closest animal relatives is given. The erect bipedalism, freeing hands from locomotor duties, seems to have appeared first in the process of evolution leading to humans, several million years ago. The causes are not clear – they may be wading in shallow waters on the edges of inland lakes and rivers or chasing prey animals across savannahs. The appearance of the ability to produce sounds of articulate speech – lowering the position of the larynx in the throat – appeared next, though it is difficult to determine when. This ability to produce sounds as units of the arbitrary symbolic communication opened up great possibilities for communication and exchange of technologically and socially relevant information that facilitated the management of environment by humans.

**Keywords:** Erect bipedalism, Future, Genetic engineering, Language, Stone Age, Tools.

It all began four or six million years ago with a walk upright on the ground. It was not the first time a tree dweller had descended to the ground and it would not be the last but, thankfully for us, it developed into a habit. We became bipedal about five million years ago [1, 2]. However, this development may have been simply related to wading through swamps [3, 4], and picking food from bushes and low branches. Despite *Australopithecus afarensis* possibly having been a good erect bipedalist three million years ago, it is now thought our ancestors did not venture far from their wooded environments until about a million years later [5].

Some time was to pass before *Homo sapiens* trod the Earth, for a few new obstacles needed to be overcome; problems created by this new form of locomotion - erect bipedalism.

Walking erect on two legs necessitated a number of changes to body structure. Not the least of which involved procreation. Four legged-animals tend to have wider hips than erect bipedal mammals which must have their legs to be brought close together for efficient walking and running [6]. That consequently meant a narrower pelvic structure, which in turn meant more difficult birthing. Over time, evolution took care of the problem by ensuring human babies were born earlier in their development and with soft heads, which made exiting the womb easier and less dangerous for both mother and child. As the result, our offspring are virtually born premature (altricial) with wider and more mobile infant skull plate joints, an advantage if the brain is to grow substantially [7].

By about two million years ago, our ancestors had survived the added pitfalls of living in the open to develop an incipient technology and more sophisticated social structures. They learned to fashion crude tools – principally a variety of wooden, bone and antler objects supplemented by sharp-edged rocks, which were to remain at the pinnacle of technological accomplishment for hundreds of millennia [8]. A simple invention, it brought many extra social benefits, through use in hammering, cutting, skinning, whittling sticks fitted with stone spearheads, and so forth. It also contributed to the beginnings of art and religion: the former *via* carving and the latter in ritual burial and ceremonial purposes.

At some time in our early history, there came another breakthrough which made us really human. This was largely due to relocation of one of the body's organs, the larynx or voice-box [9]. In chimps and other animals, the larynx is positioned much higher than in our throats.

In fact, chimps then and now, breathe through their noses while they eat: we cannot (which helps explain why thousands of us die every year by choking when something slips past a little slap of cartilage that automatically diverts air and food between our lungs and gullet and "goes down the wrong way"). Also, because of the voice box's location high in the throat, chimps and our ancestors could only grunt, a not very expressive means of communication.

With our voice-box located lower in the pharynx we achieve a large resonant cavity above the vocal chords and below the back of the tongue. This produces clear sounds of the vowels necessary to separate consonants in the syllables that are basic units of our spoken language. Having already acquired a primitive symbolic language (grunts and gestures) humans began to develop a more sophisticated form of communication through speech (words) [10]. Cultural evolution went into overdrive. Communication based on symbolic language enabled the rapid exchange of information and inter-generational knowledge transmission. However, sharp stones were still leading-edge technology, while humans thrived.

How humanity endured natural catastrophes such as Ice Ages, comet impact and climate change is pure speculation but there is a strong train of thought suggesting the survivors used their precious gift of language well: they co-operated, exchanging information necessary to locate and use dwindling food sources, and by sharing resources - and in doing so possibly developed even more altruistic attitudes towards one another. Not only did humans survive that Stone Age – they flourished.

The exchange of information has assisted in the maintenance and growth of knowledge ever since. Over the millennia, language has become more sophisticated, eventually recorded in books, stored in libraries and computers, and it is no longer inevitable that some data will be lost with the demise of a generation.

Today, we have medicine and booming technology. Problem genes, once restricted in influence, will become more and more widespread as manipulation – no matter the motivation – replaces aspects of natural selection [11, 12]. Genetic engineering raises massive possibilities and poses new questions, for example: will future generations continue to opt for designer babies? Will parents determine the eye and skin colours of their children as they determine sex today? Will genetic engineering mean the end of cosmetic surgery? Will gene tampering enable youthful life spans to reach the double century? Will our kids grow smarter and smarter until the world is full of little geniuses? Various futuristic authors such as Ray Kurzweil and Nick Bostrom maintain that humans will supersede their bodily limitations eventually becoming 'post-humans' – biologically enhanced human beings.

## AT THE END OF OUR JOURNEY, SO FAR, WHAT HAVE WE LEARNT ABOUT BEING HUMAN?

Genetically speaking, there is just about one per cent difference between us and chimpanzees [13]. Strip away all that technology and we are simply descendants of our Stone Age ancestors. For the first time in the journey, the choice is ours. We now hold power over our own biology and environment. Are we to become *Homo super-sapiens,* or *H. extinctus?* Brain physiology is what makes us human, and this can be easily altered by drugs and medication [14 - 17]. Will our brains save us from ourselves?

# CHAPTER 3

# The First Homo

**Abstract:** A view of human evolution as a continuous process occurring over a few million years is presented. In contrast to theories replicating biblical events of abrupt creation of separate species that remain largely unchanged after they have been produced, the evolutionary emergence of humans is presented here as an extended over time process running at varying speeds in various periods, but being largely continuous from generation to generation. It is briefly documented by mention of key fossils that informed studies of human origins.

**Keywords:** *Ardipithecus*, *Australopithecus*, Dmanisi, Fossils, *Homo erectus*, *Homo sapiens*, Neandertals, *Orrorin*, Piltdown, Species.

Evolution is commonly seen as "emergence" of new species. Once a species emerges, it is complete and remains the same until it becomes extinct or produces a "new species". Such a view of evolution is wrong, it has simply replaced a 'metaphysical creator' with "natural forces". What these forces are, cannot be explained in such a scenario. Critics of the science of evolution say that the forces producing new species are random. The real story of how the living world changes is different. It relies on the observation that individuals that we arbitrarily group into a species are actually not identical, they are variable in their biological characteristics: size, colour, strength, behaviour, *etc.* Each new generation of individuals may differ slightly from others, and if these differences make them survive better or produce more offspring, the variants allowing such better survival and better continuation of their family line through more offspring will increase in frequency. Considered generation-by-generation the evolutionary process is quite subtle and rather boring.

Maciej Henneberg & Arthur Saniotis

It is only when the observer moves away from the detail and allows more time, in terms of numerous generations to pass, that a change may seem dramatic – a transformation of characteristics of one species into those of another. Such step-by-step processes are the only way for life to continue, though they may run at various speeds – from just a few scores of generations to millions – depending on pressures coming from changing or stable environments and from interactions between various organisms.

For two million years we have been members of the same evolving species: a lineage [18, 19]. We have changed in response to our changing environments, to changes in our ways of doing things (especially by using weapons and tools), and by organising ourselves into societies of ever-increasing complexity.

If you took the body of research gleaned from the plethora of fossils found in Asia, Africa and Europe - a collection spanning more than five million years - and interpreted it having regard to and understanding of the mechanisms of evolution occurring in the gene pool, a parsimonious – simple - explanation was readily found: we became human gradually [18, 21, 22].

The story we have to tell is not the story of a bitter struggle of one human species coming out of nowhere and fighting against another, but a story of learning, social cooperation and the transformation of our surroundings to better serve our needs. That story is yet unfinished: we are still evolving and will continue to evolve. This is counter to views of various scientists who have suggested that evolution in humans has come to a halt. Such a view is untenable.

In the present day, evolutionary pressures have changed. Now, they mostly come from within the human system, from technologies and social structures, governments and politics. The result has been an increase in human physiological and anatomical variability: we are more different as individuals than ever before. This has necessitated changes in attitudes, requiring more health care and intervention into the ways we live.

Understanding this constant exchange between human biology and culture is the crux of our survival.

## HUMAN ANCESTORS

The search for human ancestors is over 150 years old. The first finds were accidental: Neandertals, so called because the first announced find was made in the New Man's Valley near Düsseldorf in Germany (Greek *Nea* for new and *ander* for a man plus German *tal* [or old spelling *thal*] for a valley). With the dissemination of the Darwinian theory of evolution, the purposeful search for human evolutionary ancestors had begun. By the end of the 19th century [1891-93], *Homo erectus* remains were discovered in Java. Two decades later supposedly ancient human remains were discovered in Piltdown, England. These eventually were proven fake, but remains of *Homo erectus* are still being found in several sites in Java, China, in Dmanisi, Republic of Georgia in Europe and in numerous sites in East Africa. These date back between 1.9 Ma (millions of years) and about 400 ka (thousands of years). *Homo erectus*, however old, were not in a strict sense our ancestors, they were earlier humans of our own kind. Their brain size and body size range overlapped within the range of modern humans, they purposely manufactured stone weapons and tools of standardized, aesthetically pleasing symmetric shapes and used fire for cooking. Opinions of anthropologists are divided: some argue that forms called *Homo erectus* were really earlier members of our own species and should thus be called *Homo sapiens*, many others consider them a separate species or even argue that these remains represent several different species. Be it as it may, they were members of the genus *Homo*.

The discovery of the first ancestor of *Homo*, came during Christmas holiday in 1924 in South Africa. There, a young anatomist, Raymond Arthur Dart identified a fossil retrieved during limestone quarrying in Taung, as a form intermediate between apes and humans. He named his find Southern Ape from Africa (*Australopithecus africanus*). This first fossil was that of a few years old child with human-like teeth and small braincase. Beginning in 1936 numerous fossils of adult australopithecines were found in various sites in South Africa and since 1959 in East Africa. Today, many australopithecine fossils are still being found in Ethiopia, Kenya, Tanzania, Chad and South Africa. Australopithecines lived between 3.5 Ma and some 1.5 Ma. It is still unclear whether any fossils recovered from that time range in China and Europe represent australopithecines. Bones of australopithecines indicate that their normal body position was erect, bipedal, the same as our own, their teeth were similar to ours, without a protruding fang-like canine. Their bodies were, however, smaller than ours -- some 1.2-1.5 m tall – with proportionally smaller braincases.

Recent years brought discoveries of even earlier than australopithecine forms: *Orrorin tugenensis* from Kenya, whose limb bones indicate upright bipedal stance at some 6 Ma and *Ardipithecus ramidus* at 4.4 Ma from Ethiopia who was bipedal despite living in forests.

There is a claim that the earliest human ancestor, *Sahelanthropus tchadensis*, lived as early as 7 Ma ago. This, however, is based on geological dating (context of geological strata and faunal remains), that has been recently challenged by an assertion that the skeletal remains of this hominid form were purposely displaced and buried by medieval Muslim settlers in an "alien" geological context [23].

The earliest fossils of hominin are securely dated at about 5 Ma ago. They are teeth and bone fragments found in East Africa (Kenya, Ethiopia, Tanzania) which belonged to *Ardipithecus kadabba, Ardipithecus ramidus* and *Australopithecus anamensis*. Fossils of two other species of australopithecus, dating at 4-2 Ma ago are commonly found in Ethiopia, Kenya and Tanzania (*A. afarensis)* and in South Africa ( *A. africanus)*. In 1999, a new species *A. garhi* dated at about 2.5 Ma was described from Ethiopia, while in 2010, yet another species dated at about 2 Ma, *Australopithecus sediba* was discovered in South Africa. *A. sediba* has been subject to scientific scrutiny due to its more human-like hand; some scientists supporting *A. sediba* as a likely human ancestor. There is a possibility, as yet unproven, that some australopithecines lived in China some 2.5 Ma ago [24].

The oldest fossils attributed to our own genus *Homo, (H. habilis)* come from East and South Africa and are dated at about 2.5-2.0 Ma. In 2015 one of the oldest finds of the genus *Homo* was reported from the Dinaledi Chamber in a cave system in Gauteng, South Africa near Johannesburg [25]. The *Homo naledi* remains are dated at some 2.5 Ma. Inside an isolated chamber some 30 m deep underground, and over 80 metres away from the dolomitic cave entrance, fossils of numerous individuals (thus far about 15) of age ranging from infants to adults were found. The chamber is accessed by narrow, twisted passages and it seems human bodies were deposited there purposely, for burial by people using some form of artificial light (fire?). The exploration of this site is ongoing.

Different other finds from Africa living at about 1.7 Ma have been variously classified as *H. ergaster and H. rudolfensis* in addition to *H. habilis.* The finds from Dmanisi (Republic of Georgia, Europe) dated at about 1.7 Ma indicate undisputed presence of *Homo,* most probably *H. erectus* (though also claimed to be a separate species *H. georgicus),* in Eurasia. Some finds from Java (*e.g.* Sangiran) that are undoubted *H. erectus* could be as early as 1.9-1.7 Ma ago.

At the same time (2.5-1.2 Ma) are dated finds of robust australopithecines ( *A.* [*Paranthropus*] *robustus* and *A* [*P*]. *boisei)* often found at the same sites as remains of *Homo.* The name "robust australopithecines" derives from the fact that individuals so classified had large teeth and thus robustly built faces. Their body size was similar to early *Homo.*

Major sites where hominin remains were found, with approximate dates, are shown in Fig. (**1**), and their various genus and species names are listed in Table **1**.

Table 1. List of hominin fossil taxa mentioned in this book. By no means it is a full listing. The full list of formally acceptable hominin fossil taxa is close to 60 [36] with new added nearly every year.

| Dates in millions of years | Genus | Species |
|---|---|---|
| 2.5-0.0 | *Homo* | *antecessor, erectus, ergaster, georgicus, habilis, naledi, neanderthalensis, rudolfensis, sapiens (floresiensis)* |
| 4.0-1.2 | Australopithecus | *afarensis, africanus, anamensis, garhi, sediba, aethiopicus [Paranthropus]robustus, [Paranthropus] boisei[Kenyanthropus] platyops* |
| 6.0-4.0 | *Ardipithecus* | *kadabba, ramidus* |
| 7.0-5.0 | *Orrorin* *Sahelanthropus* | *tugenensis* *tchadensis* |

## THE ORIGINS OF MODERN HUMANS

There are two levels at which the problem of the origin of modern humans can be considered:

1. The origin of our own species *Homo sapiens*.
2. The appearance of people who acted practically identical with ourselves.

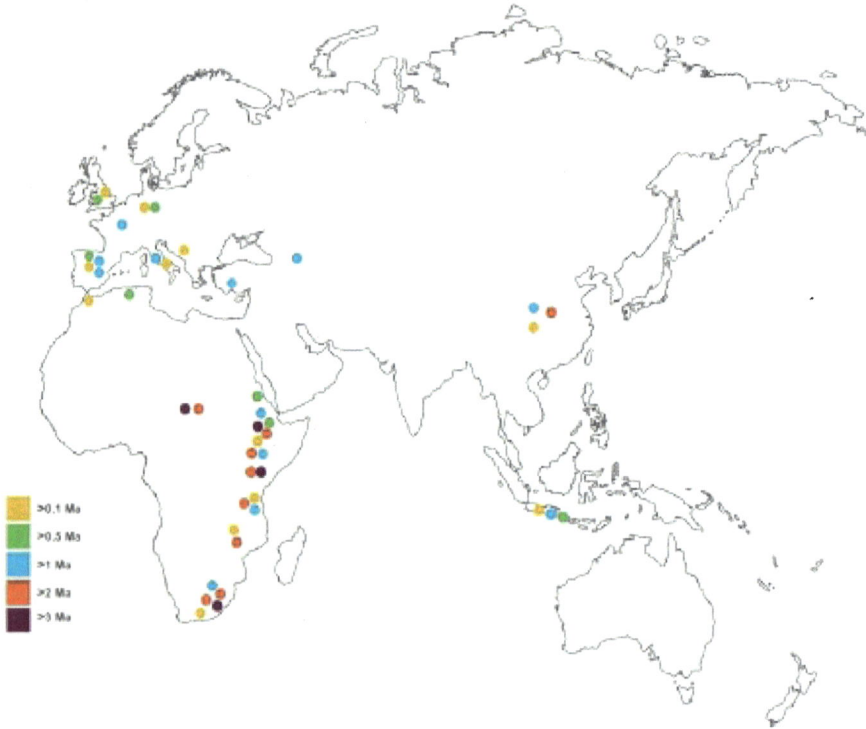

**Fig. (1).** Sites where fossils of early humans were found. The distribution of the sites is a joint result of the presence of hominins in the past, physical conditions for their preservation and the intensity of searches that depends on funding and ease with which fieldwork can be organised in various countries.

These two levels are not identical. The first one asks a question about the origin of those humans who no longer can be classified as *Homo erectus*, but not necessarily are indistinguishable from ourselves. For example, a number of human remains were found in Spain dating as far back as 1.2 Ma, in sites such as Fuente Nueva 3 in Orce [26], Sima del Elefante, Sima de los Huesos and Gran Dolina in Sierra de Atapuerca [27 - 29]. Some of them were called *Homo antecessor* (ancestral human). They show traits intermediate between earlier *Homo erectus* and later neandertals and, eventually, modern humans. The period between, roughly 400 ka and 100 ka is characterised by a number of hominin fossils who look more "advanced" than *H. erectus*, but are not as "progressive" (?) as we are. Some of such forms (*e.g.* neandertals) survived until 30 ka ago.

Some anthropologists distinguish between "archaic *Homo sapiens"* (400-100 ka), "neandertals" (150-30 ka) and "anatomically modern *Homo sapiens"* (100-0ka ago). The main difference between "anatomically modern" humans and the other forms is the robusticity of the skeleton, especially of the face. Body size, and thus cranial capacity of all forms was roughly the same as of today humans (about 1.5-1.8 m, 900-1800 ml). Sometimes, like in the neandertal group, the average cranial capacity was slightly greater than today (1400 *vs.* 1350 ml) due to robust body size.

The neandertals are a much disputed group [30]. They lived at the time (approx. 150-30 ka ago) from which are also known remains of "anatomically modern" humans. The neandertal skeleton was characterised by wider pelvis and thicker bones, large forward protruding face with prominent nose, big sinuses, heavy brow ridges, and less pronounced chin. Their body height and brain size, however, were about the same as of modern people [31]. Apart from creating a range of stone and wooden weapons and tools, Neandertals produced sophisticated artifacts (stone, bone and wood, jewellery) and seemed to have had complex social structure (graves decorated with ochre paint and flowers, support for disabled people). Their remains are sometimes found at the same sites or in the same areas as more gracile "modern" people (Palestine, Romania, France). There are two scenarios of the ultimate fate of neandertals favoured by various researchers: (1) "Brutal" neandertals were a separate species that our own more sophisticated ancestors coming out of Africa killed ruthlessly ("replacement hypothesis") and (2) neandertals, being a variant of our own species adapted to extreme cold of the glacial era, simply mixed with more gracile forms and evolved into ourselves. Recent studies of genomes of neandertals and of their contemporaries - "denisovans" - show that present-day people share some genes with those earlier forms [32, 33].

Although the fossil and genetic evidence for human evolution is abundant, it is still incomplete. The question of our origins is obviously an emotionally loaded one. Therefore, the various hypotheses constructed to explain modern human evolution are coloured by historical experience of their originators. For example, there are two schools of thought about the origin of anatomically modern humans. One school argues that the "moderns" arose in one spot, in Africa, and then expanded over the rest of the world pushing aside, if not exterminating, native peoples living there. The other school maintains that people living on all continents evolved together towards modern form as conditions changed. The first school is labelled, for short, "out of Africa", the other "multiregional". The "out of Africa" school is intellectually influenced by the experience of Northwestern European colonialism of the last 400 years. In this experience British, and to a lesser extent Dutch and French, people invaded other continents and islands and decimated natives who were considered "indigenous" inferior races. Colonial activities of Southern Europeans, especially Portuguese and Spanish and of Russians were somewhat different. Though colonialists fought natives, eventually they mixed with them producing blended populations. Many other nations of Asia, Europe and Africa, though invading each other, generally intermixed and blended their gene pools and cultures. This is closer to the "multiregional" model.

Despite all these theoretical differences and uncertainties, one thing is quite sure: there are numerous fossils of forms that can be labelled nothing else but "human" dating back at least 1.7 Ma that were found in Africa, Asia and Europe. Their ancestors in Africa, and possibly in Asia, go back to 6 Ma. Older fossils look less like ourselves, younger fossils are more similar to us. Therefore, an evolutionary process gradually transformed beings more similar to apes (but not identical with modern apes) into people that today live around us. This process was by no means smooth - there could have been periods of slower and of faster change - and certainly not directed towards some predetermined "progressive" ideal. For instance, although the size of the hominid brain had been increasing from about 3 Ma ago to about 100 ka, in the last 10 ka the volume of human brain all over the world decreased by about 10 % (from approx. 1500 ml to about 1350 ml). These changes were not related to "intelligence" but simply reflected changing body size [34, 35].

# CHAPTER 4

# The Problem of Species and its Handling by Anthropologists

**Abstract:** The unclear concept of "species" as a category of biological research is discussed as specifically applied to human evolution. The concept is a rigid 18th century pre-evolutionary category that complicates biological thinking about dynamic processes of evolution. Currently, there are 23 defintions of "species" used by biologists. With the above in mind, a discussion of the hominin fossil record is presented in terms of the number of purported "species" into which the fossils can be straightjacketed. The conclusion is that the simplest hypothesis, that of a single human species being present at any point in time during the last few million years, cannot be reliably falsified with the evidence currently available.

**Keywords:** Ancient DNA, Darwin, Diversity, Linnaeus, Mayr, Palaeoanthropology, Species definition, Variability.

What's in a word? When it comes to science, every one matters, and in the world of anthropology the most contentious word of all is 'species'.

As defined in the 18th century by Carl von Linné, the role of the scientist was to understand the order of nature as created by God, by identifying the types of animals and plants which had been created. Each was regarded as being an ideal member of a species created separately for its own continued, unchanging existence. This was a classic example of 'biological reductionism' which informs the life sciences.

Charles Darwin, when proposing his theory of evolution had to refer to that concept, because at that time there was no other method of classifying plants and animals. Even then he had trouble defining 'species'.

**Maciej Henneberg & Arthur Saniotis**

He stated in his seminal book on the *Origin of Species* (1859), that the very concept was indefinable – meaning not really scientific. He clearly stated he did not see a problem with the notion that some variations could occur within one species.

## THERE IS STILL NO AGREEMENT TODAY

In current biological literature, there are more than 23 different definitions of 'species' [36, 37]. They roughly fall into two categories: Typological, based on platonic essentialism – the definitions accepted by creationists; and Biological – evolutionary concepts. There are variations on the theoretical themes and hybrids of both.

The classic essentialist typological definition is that a species is the ideal type as defined by God and all individual members of a species are just special instances of one type. To illustrate, look at the Blue Warbler. It appears to be a copy of the ideal type because it is small, has blue stripes on its wings, a beak, feathers, *etc.* Several of them flocking together display variations in size, markings and so forth but while there are a few departures from the ideal, all Blue Warblers look generally similar. The typological definition of a species is, basically, that there must be uniformity of its members' physical characteristics.

However, the typological concept does not explain mutants and apparent deformities. It does not recognise such variants as playing any significant role and dismisses them as mere distractions from the ideal type. This typological concept is still adhered to by some leading scientists.

The biological concept was introduced by Ernst Mayr, a German-born American zoologist at Harvard University, who described a species as a conglomerate of individuals sharing a gene pool – in other words, individuals capable of exchanging genes with one another. There must be a mother and a father capable of producing progeny and together those individuals constitute a species which is processual, not typological.

The biological dynamics of such individuals include living together, sharing the same environment and interacting reproductively - not necessarily sexually, because some may produce offspring by parthenogenesis. Production of offspring is a commonly recommended test in defining a species.

A population, members of which are exchanging genes with each other from one generation to another, might frequently encounter in its travels environmental barriers such as rivers, mountain ranges or oceans. Some individuals may cross those obstacles: many might not.

One group becomes separated and isolated from the rest and finds itself in a different environment. Some individuals previously successful at reproduction might not be as successful, while others previously unsuccessful begin to produce larger numbers of offspring. Over generations, the proportions of genes carried in the total population pool change and the new population begins to look different on average from the one left behind the barrier.

Mutations always occur in the gene pool and may be passed on more or less often through reproduction, depending on whether or not they offer some practical advantage. If they do not, they are slowly eliminated over generations because no offspring are produced to carry a gene further.

However, when a scientist attempts to give an example of a proven incidence of speciation, she runs into trouble, for there is not an indisputable one [38]. The problem with speciation is that no-one has ever witnessed it - though it is well described by the biological concept.

There are good suggestions, for example, among fish in lakes which become partitioned or isolated [39], but so far there has not been a single occurrence of definitive speciation, followed and observed in nature. It is not that evolution does not occur, as suggested by some opponents who use the lack of witnessed 'speciations' as an argument against evolution, but that species are so ill-defined that it is debatable when exactly a new species appears in the succession of generations. Species is not a scientific unit of analysis, same as we conceive an atom or a light wave. It is a descriptive term based on a false assumption that ideal templates of plants and animals were somehow established and all organisms must fit one of those templates. What actually exists are variable individuals who may be more similar to others than to anyone else, but it is up to imprecise human impressions to group those individuals into collective units – simila [40]. Such collective units are ephemeral – they change as generations follow each other. No species can last forever because conditions for its existence change through time.

The question arises as to whether or not neandertals and other early hominins are simply variations of the human species, but since it is impossible to get fossils to reproduce, scientific judgement must be based on secondary characteristics.

For example, researchers might say that neandertals and modern humans could not have had sexual relationships and produced children because they could not speak, and therefore, they could not have communicated adequately enough to form stable social unions. But whether or not neandertals could speak is a contention based on the study of certain anatomical characteristics which either prevent or allow speech - and often such research can be interpreted one way or the other.

Studying the past can never impose the ultimate test of the biological concept - the ability to produce viable, fertile offspring. Science must rely on deductions often several steps removed from the actual act of procreation.

That is not to say there has never been another hominin constituting a separate species. However, the fossil record does not offer incontrovertible proof that other hominids which were unable to produce offspring capable of mating with our direct ancestors ever existed – so, we simply don't know.

In science education, if you don't know you should not speculate. Good science demands acceptance of the simplest explanation – the Null Hypothesis, which is: our ancestors have always belonged to one single species, because so far there is no conclusive proof that they didn't [22, 35].

There is a paradox related to the identification of our ancestors. The only criterion to say that a particular fossil is our ancestor is that it is similar to us. This, if applied literally, would prevent us identifying any fossils that do not look as ourselves as our ancestors. In order to be more inclusive, we postulate that our ancestors were at earlier stages of processes that produced us. For example, we assume that our brain size increased through time. Thus, if we project the curve of increase backwards, we can identify those smaller-brained organisms that fit the assumed process. This is illustrated by a Fig. (**4**) from De Miguel and Henneberg (2001) [41] redrawn here as Fig. (**1**).

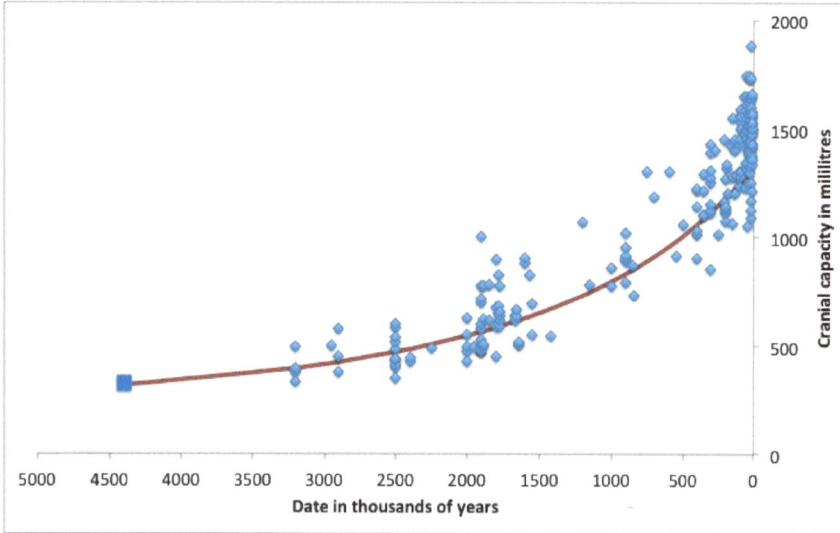

**Fig. (1).** Being a Fig. (**4**) from De Miguel and Henneberg (2001) redrawn to illustrate the processual identification of ancestors. Data scattered are cranial capacities of hominins. The double exponential curve is fitted to those data. It runs back in time to 4400 ka where it hits the cranial capacity of *Ardipithecus ramidus* (large square). This indicates that *A. ramidus* could be our ancestor.

The same applies to the shape and size of teeth, face and the whole body. Thus, we search for fossils that fit our perception of how our own lineage changed through time, while at the same time we want to find proofs that, as is a general regularity in evolution, earlier forms underwent adaptive radiations producing numerous species. We can hardly identify those species that radiated far away from our own lineage if we use the idea of processes occurring in our own lineage. Were one of the species descending from our ancestor 6 Ma ago reducing its body size and thus brain size while developing larger canines and changing its mode of locomotion from erect bipedalism to suspensory brachiation, we would have little chance of identifying it in the fossil record.

Thus far, paleoanthropologists argue for only one possible radiation away from the human lineage. This is the development of "robust" Australopithecines – organisms that, living in parallel with early Homo had smaller brains and bigger teeth than these latter. Their body size was comparable with that of earlier Australopithecines. This distinction may, however, be a case of wishful thinking because "robusts" seemed to have very small range of variation in brain size and in tooth size. It may be, that the entire range of variation of hominins at about 1.5 Ma ago has been artificially divided into more "progressive" *Homo* being closer to the human "ideal" of bigger brain and smaller teeth, and more deviated ("derived") robusts with less human brain size and tooth size. When all brains of that time period and all teeth are taken together, their ranges of variation are not significantly different from those of a single modern human species.

Another instance of reconciliation of the paradox is an attempt at presenting neandertals as very recent evolutionary side branch of human evolution. This human race, with powerfully muscled bodies, correspondingly large brains and a complex technological and cultural endowment is said to be a species separate from ourselves though practically every characteristic of their bodies overlaps with the range seen in our own species, however sometimes lying away from the mean. During the Ice Age, neandertals adapted to life in harsh conditions of cold climate in which survival depended on the ability to hunt large animals such as mammoths, woolly rhinoceroses and bison-like ungulates. Very small fragments of DNA possibly preserved in neandertal bones, show some differences from the average modern human genome, but such differences are not impossible to occur as a joint result of post-mortem DNA degradation over some 50,000 years and expected evolutionary change over same period of time. Researchers trying to extract and analyse neandertal DNA automatically assume that any modern-like DNA extracted from neandertal bones is a result of contamination with DNA of excavators, and accept only DNA fragments differing from modern DNA as true "neandertal" one. This is obvious wishful thinking. Most neandertal bones yield only modern-like DNA, the "non-modern" being detectable in less than 20% of neandertal bones [42]. Recently, cases of 20-30 ka old skeletons having a mixture of "modern" and neandertal traits were described from Lagar Velho in Portugal [43] Pestera Mulieri [44] and Cioclovina [45] in Roumania. The need to support the principle that evolution progresses by creating new species leads to the re-labelling of a particular human local population as a separate species. Such approach is dangerous because it produces a precedent by which some currently living human population may be labelled a separate species.

The most recent example of the attempt at proving the evolutionary point of multiple species emerging from a common ancestor is the creation of a species of *Homo floresiensis* out of a limited assemblage of human remains including only one single skull found in Liang Bua cave on Flores island, Indonesia. The skull is dated at barely 17 thousand years ago, while other remains fit into a range of 95-12 ka, widely overlapping with the time (some 40 ka) when modern humans lived in Indonesia. The skeletal material attributed to *H. floresiensis* may well derive from a local population of small-bodied (pygmy) modern people, some of whom were afflicted with growth disorders, not uncommon among modern people [46 - 49].

## HOW SPECIES ARE CREATED?

The appearance of a new species is an accident. Life that maintains a particular informational structure through metabolic turnover of chemical components tends to preserve existing body forms and functions. Alas, due to mutational propensity of the genetic code, organisms are constantly exposed to alterations. Sexual reproduction, a common way to cover up mistakes in the code, requires exchange of genetic material between individuals.

Mutations of specific single genes are rare, individuals are often dispersed in the environment. A few mutations occurring in some individuals who happen to live at a distance preventing mating over a number of generations may result in such incompatibility of the genetic material of some groups that they will be unable to mate effectively since disparities between their alleles are too great. In this way two, or possibly more, groups of individuals who are reproductively isolated from each other arise. This reproductive isolation precludes "correction" of differences in genetic codes of the two groups. Natural selection will differentially affect these two gene pools either because different genetic properties require of individuals different approaches to similar environmental conditions or because the two groups live in different environments.

Once the way to mix, and thus equalise, genetic information is closed, each group of individuals experiencing further different mutations and adapting in their own ways to their respective environments will arise. Instead of having one kind of life, one set of repeatable information, there will be two, or more sets each trying to continue its existence in its own way. Two kinds of life will exist instead of one. Reproductive isolation of individuals who slightly differ in their genetic endowment does not necessarily require isolation in space. It may occur due to differences in body size, colouration or behaviour that will render some individuals reproductively unattractive to some other ones even if they occupy the same territory.

Although freshly separated groups of individuals usually do not compete with each other directly, as they have different environmental requirements and often can be separated in space from each other, ultimately they become competitors instead of being mates. This happens because without exchange of genetic material, the existence of each group becomes useless to the other one while both groups must use resources of the same Earth. It is true that some animals eat plants in order to survive, some plants eat animals for the same reason, while other animals live in the bodies of plants or animals as parasites. Some plants and fungi even form inseparable alliances growing as lichens. Such things came later in the history of life. Initially, separations of groups of individuals were sad events limiting and fragmenting gene pools.

The first people looking at the diversity of living forms could hardly comprehend it. In most cultures they tried to cope with describing this diversity for purposes of hunting, gathering and later, agriculture by grouping plants and animals according to some common, easily observable and useful characteristics. For example, plants bearing plump juicy fruit were called collectively "fruit trees". Although, technically, a walnut tree produces a fruit it will not be commonly included into the "fruit tree" category because its fruit's flesh is very bitter, awful tasting and stains human skin. The stone of this fruit, however, contains a tasty kernel - a nut. We all know that animals living in water are fish, so we have shellfish, crayfish and even, according to some TV commentators - a whale fish. In fact, the Dutch word for a whale in the 17th century was *walvis* (whale fish precisely). There are people who call a dolphin "this fish". Modern biologists will tell you that these kinds of "fish" are not really fish at all, since shellfish animals are really molluscs akin to garden snails, that crayfish is phylogenetically related to a spider rather than to a shark, and that whales and dolphins are really like us - they are mammals. All this is because scientists looked closer at traits that are not so obvious, but which indicate common ancestry. Basic characters of body plan (phylum) or functions are considered here rather than common environment or similar practical utility.

All this notwithstanding, scientific classification is just a bit of an improvement on what people of all cultures did for thousands of years - assigning individual plants and animals to "kinds" created on grounds of the similarity of certain characteristics. Both common people and scientists wished to explain those groupings by finding some general regularity. In her famous essay on food categories in the book of Leviticus in the Old Testament, the anthropologist Mary Douglas speculated why some animals were *kosher* and others were not. She noted that *non-kosher* animals were really anomalous due to their contradictory physical semblance which precluded them from neatly fitting in a clear category. According to Douglas, the demeaned pig, is a classic anomaly – having cloven feet and suckling its young, but does not chew the cud like cattle and sheep. Therefore, pig = NON-KOSHER. Another animal which is omitted from ancient Hebrew cuisine is the winged anomaly – the bat. While bats can fly like birds they lack feathers, nor do they lay eggs, but like other mammals give birth and breastfeed their babies.

The easiest explanation was to say that every organism was somehow created. Whether the creating force was assigned to a higher spirit, or to inanimate objects, is of lesser relevance. Medieval people believed that mice arose from the mixture of dirt and excrement. The problem with creation of various kinds is that all individuals of the same kind should look and behave the same. Alas, they do not. Some ducks are larger than others; moths have different colours and patterns on their wings *etc*. Each kind can be subdivided into special categories - "species". The problem does not end here because each species contains individuals that are not identical.

The important characteristics of a species is that its members can mate with each other to produce viable and fertile offspring. Thus, a species has common gene pool. This pool, however, as we already described, can become subdivided.

Besides, members of some different species can mate and produce viable offspring. For example, horses and donkeys, or lions and tigers. Some offspring of horse/donkey mating will even be fertile. So will be the offspring of a wolf and a dog. Although, wolves and dogs can be considered the same species according to the biological criterion of interbreeding it is practical to keep them apart as separate kinds because of their different roles with respect to humans and human economy.

As individuals of a given species reproduce themselves from one generation to another they will change due to mutations and natural selection. If we wait a sufficient number of generations the change may be such that the descendants will be quite different from ancestors. Brought together by a time machine they may even fail to mate. Will they still be members of the same species?

Some palaeontologists argue that not all dinosaurs died off some 65 million years ago. Some dinosaurs, they say, became birds already more than 100 million years ago. These were small carnivorous dinosaurs similar to *Velociraptor* - creatures running on two hind legs along the ground and using outstretched arms to capture their prey. If this is true then there must have been an incessant series of parents and offspring leading from a Cretaceous scaly dinosaur to our pet budgie in the cage in the living room. Yet when we eat chicken we do not feel like consuming a dinosaur. Are they the same or not?

The answer depends on how we think. Obviously the chicken on our plate was born a few months ago while some ancestral dinosaur died at least 65 million years ago. Assuming that dinosaurs, and later birds, lay eggs only once a year, we can postulate that there were 65 million steps from parent to offspring that separate dinosaurs from a roast on our plate. The point is, given enough time mutations and natural selection can change a lot of things. They are the same in the sense that chickens contain many genes of original dinosaurs, but also they have enough genetic material altered by mutations and amplified by selection to be somewhat different. They are therefore partly the same and partly not. If we, however, believe that there is such a general, eternal thing like a "model chicken", and that there were "model dinosaurs" we will categorically answer "no" to the question "are they the same?"

The "model chicken" and the "model dinosaur" exist only in human heads. They are products of our reasoning that tends to put things into neatly divided categories. We like to see things that are either black or white, good or bad. We classify people and things in order to apply to them standard behaviours. Mothers advise their children "do not talk to strangers". They are disappointed when children do talk to some people. Why? Because children do not see these people as strangers, while parents do. When some of us see a black person approaching our car we often assume that he will try to attack us or steal something. How embarrassed we are if he turns out to be a new professor on our faculty.

Kinds of animals and plants, or their "species" exist only in human heads. In reality there exist only individuals. Depending on their genetic code some of those individuals will be able to mate with some other ones and produce offspring who in turn will pass their genes to their own children. Two successfully mating individuals do not have to be very similar - take various breeds of dogs for example - a Dachshund and a Maltese. Usually mating individuals will be similar because they must share substantial part of their genetic code to be able to mate, there is, however, no special limit to the number of genes that may differ.

Individuals, mating, new individuals, mating again, yet new individuals and so on and so on is all that really exists. The categories into which we try to squeeze individual plants or mammals are the constructs of our minds created for understanding our complex world. Our understanding need not be correct.

Mating individuals and their offspring can be arranged into long lines of ancestors-descendants-descendants of descendants-descendants of these descendants *etc.* These are called lineages. Members of the same lineage share the same genetic material, but this does not mean that this material is identical between generations. It is a bit like with the old mansion that belonged to the same family for several centuries. Some extensions were added, some rooms removed, some others joined to form a large hall. The building is not the same as it was 300 years ago and yet it is the same mansion. We can say that the mansion evolved over the years.

The history of humans is the history of a particular lineage. We know that this lineage existed because we are here with all of our DNA being mapped out. This DNA must have come from our ancestors. How exactly did all this happen is still a bit unclear. We must descend back along the human lineage, along the family tree of each one of us. Only a small part of this genealogical record has been well preserved. Most is just bits and pieces of fossilised bone and teeth.

### *H. erectus* & The Hobbit

There's another problem, in that a past species member, different from today's *Homo sapiens*, may be different just because it is from the past and not from the present: it may have evolved through different environments.

That is not a majority view among scientists in general or anthropologists in particular. The counter-view is justified by arguments that some differences observed between various ancestral hominins are large enough to conclude that they must belong to a different species. For example, brain size is often used when making such comparisons [50].

To some scientists, *H. erectus,* who lived some 1.9 – 0.6 Ma ago, is a different species –but not to all [51, 52], An example of statistical difference is that the average brain size of modern humans is approximately 1350 millilitres with a range from 880 to 1850 ml. The average brain size of a group of *H. erectus* from one million years ago was about 900 ml. It is a matter of judgement to conclude whether or not the average difference of 400 ml is enough to define a new species. There are other characteristics: the tooth size of *H. erectus* is bigger on average. In relation to brain and teeth sizes there is an overlap in the full ranges of both, and there are some modern humans with brain sizes smaller than some *H. erectus*.

Human physical characteristics continue to alter historically.

Language we use to discuss evolution poses a major difficulty, because language evolved in a society that did not yet understand evolution, and it is not suited for modern discussion of evolution.

It is not clear for anyone how much difference between individuals must there be to include them into separate species. We are talking degrees of separation. How big must teeth be to differ from another individual's to the extent that we can say it was from a different species? Sometimes interpretation of statistics is just a matter of opinion, though we can also use elements of lifestyle gleaned from the archaeological record.

At least 80% of biological anthropologists, if not more, adhere to the multi-species theory of human evolution, and would be happy to support the Hobbit as a new species if it was proven to their satisfaction, compared to less than 20% of biological anthropologists who believe that there are not many species in the human lineage.

There's a raging international controversy over the exact nature of the Hobbit, the diminutive being labelled *Homo floresiensis* whose remains were found on the Indonesian island of Flores in 2003-4. If the Hobbit is to qualify as a new species – as its finders claim - and not be regarded a pathological specimen - as other scientists believe - what would constitute scientific proof? There would need to be found a population of similar individuals, not necessarily identical, and some sort of proof of their continued existence.

DNA carries genetic information which reflects changes to genetic structure, and it is a good way of defining a given population and therefore the gene pool of a species. The problem is that the DNA of one individual does not prove anything anyway. This is of major importance to the biological concept, in that a species must comprise a dynamic population of inter-acting individuals capable of reproduction over time.

Therefore, in order to fully prove that the "Hobbits" are a new species a number of skeletons of small people without any evidence of pathological condition would have to be found, but the acceptance of this new interpretation would depend on how their physical characteristics differed from our own. If their stature was within the range 1.0 to 1.5 metres there would not be enough difference between them and modern humans to decide there was a gap. The biological concept of species demands there must be a clear gap in the distribution of physical characteristics - such as the biggest individual of one species must be smaller than the smallest of another for it to constitute a different species.

The Hobbit does not qualify because the range of its reconstructed stature runs from 1.06m to 1.35m, whereas the range for pygmy members of *H. sapiens* runs from about 1.05m to 1.65m, and recent finds have been made elsewhere of small statured people [53].

Sometimes we use combined morphological features, like brain to body ratios, and if there is a gap in the ratios then suspicions might arise about a new species. The ratio in the Hobbit's case is very different, and there is an actual gap. The smallest brains of non-pathological humans are about 850 millilitres, the Hobbit's only about 430ml – so there's a gap not covered by any other normal non-pathological human. But there are pathological conditions evident in the Hobbit [40], and standard taxonomy dictates that identification of a new species cannot be based on observations of pathological conditions. And one has to be cautious with DNA, because there is no one template of human DNA, but quite a range.

Those who oppose the idea of multi-species are usually those who have had broad biological training, possess broader outlooks, and have good understanding that variation exists within each species [20, 54]. They also understand the importance of ranges of biological variation that can be tolerated within a single species. There's no agreed standard of how much variation must exist to separate one species from another. It becomes a matter of personal opinion. There is also the issue of typological thinking, as all humans are naturally inclined that way: we like to think of ourselves as separate – short, tall, fat, thin, black, white, *etc.*

Archaeology, for example, is especially sensitive to typology, because it describes human cultures. Human culture is determined by human minds that like categorical distinctions, and the archaeologist's natural way of thinking is that everything occurs in distinct types. Therefore people studying human evolutionary past whose backgrounds are in archaeology, tend to cope with a variety of earlier human forms by separating them into distinct entities – species – rather than tolerate ranges of individual variation.

One of the strongest supporters of the single species theory in human ancestry was the late Frank Livingstone [55], who was primarily a population geneticist, as is Robert B Eckhardt of the Pennsylvania State University, a geneticist and human biologist. This is a relatively new science which works very well in the area of biological species variation. There are also other people whose primary education and experience is in biological anthropology, the study of fossils, and some of them believe there are very few other species, (*e.g.* C. Loring Brace and Milford Wolpoff of the University of Michigan).

There are some who believe there is just one lineage. They have backgrounds in palaeontology, like Francis Thackeray, the former Head of the Human Origins Institute, in South Africa or in human population genetics and human variation like MH. The world renowned late Philip Tobias, whose first degree was in medicine and whose life centred on biological anthropology, would subscribe to the few species theory rather than many.

Believe it or not, there is a textbook by C. Loring Brace, University of Michigan, one of the seniors in the profession, now in his eighties, called *The Stages of Human Evolution*, and we share his view that there was a sort of linear progression in human development, starting with australopithecus grade, moving to early *Homo*, then to Late Archaic *Homo*, and to ourselves. He also included neandertals within our species – so yes, there is such a view.

Views of anthropologists on human origins are very disparate. Part of those differences result from personal choices depending on their backgrounds, the other part on the lack of uniform criteria within the discipline. There is certainly a lack of cohesion within the discipline, especially in palaeo-anthropology which has attracted involvement by geologists, physicists and members of other disciplines because it is such an exciting field. There is no universal school of thought within palaeoanthropology. This is very peculiar –what is worse, there are no clearly-defined groupings of thought within the discipline.

Darwin's theoretical explanation of human origins is still sound, though today we have many more facts and detailed interpretations. One is a Modern Synthesis, a theory of evolution which arose in the mid-20th century. It has now been expanded by advances in genetics and evolutionary interpretation of individual development of organisms.

There have been textbooks on biological anthropology written by a number of different individuals, and all authors have the responsibility to inform their students at least about the correct terminology. A palaeoanthropologist would be hesitant to write a textbook in which he/she did not mention at least 15 hominin species, because while he may challenge an interpretation, he/she would concede that students need to be kept informed of the range of views held by anthropologists, and be made aware of what various species are alleged to have existed. Plus, from the students' point of view, it is interesting to learn all those names given to our ancestors – it adds colour to the subject matter.

All courses regarding human evolution involve systematic exposition to a large body of knowledge. In many countries, a postgraduate qualification can be obtained by doing research in depth into just one aspect of anthropology - fossils, for example, from Sterkfontein in South Africa – as long as the student holds the required educational standard, a tertiary degree, pretty well in any discipline. This is the case in most British Commonwealth countries.

In Europe, the system usually requires the candidate to produce research work and pass comprehensive broadly-based examinations in biological anthropology, and in an auxiliary discipline. Traditionally, that was philosophy, but now there is a choice. The candidate must possess at least an adequate base of knowledge in related disciplines, and be prepared to make an oral defence before a panel of academics and the general public. This system is the legacy of mediaeval times, from places of higher learning in Italy, Germany, France and England. It demands the applicant prove worthy of becoming a broad-minded academic.

In the United States, generally speaking, a student is required to undertake course work and face examination before being allowed to perform research into a single topic. In Australia, there is no set standard. It was decided it was too expensive to bring in overseas examiners for oral defence.

It can be seen that contributions to research in the field are eclectic, and can be submitted by people not having any anthropological knowledge. Opinions of individuals always vary, but if researchers share a common professional core, they can at least conduct structured discussions. If a person does not have that "commonality", it is difficult to do so.

At present anybody with some academic degree who did some research in some aspects of human evolution or human variation can describe himself as an anthropologist. The minimal knowledge required of a thoroughly-rounded biological anthropologist, should include genetics, population biology, taxonomy and systematics. These are areas which must be well understood, at least at the senior undergraduate level. This might require a semester in each to qualify.

In the USA, each university has freedom to set the structure of curricula and biological subjects as well as subjects like archaeology or social anthropology are included in courses leading to degrees in biological anthropology. There are variations on the theme, but in general all those subjects need to be studied.

# CHAPTER 5

# The Phenomenon of Evolution

**Abstract:** Patterns of thought about human place in nature and organic evolution are discussed. Basic reasons for placing humans among mammals rather than considering them biologically exceptional are given. The historical origin of the, still currently used, taxonomy of living things that classifies them into distinct static categories is discussed as an impediment to the understanding of the variable and dynamic nature of life. This dynamic nature of life producing variable organisms in every generation makes evolution an inevitable process commonly occurring today same as it has occurred in the past.

**Keywords:** Aristotle, Darwin, Mars, Platonists, Schiaparelli, Socrates, Taxonomic system.

Evolution is largely perceived as a theory explaining the past events that led to the appearance of various species, but it is considered to be irrelevant to what happens to humans today. This view is erroneous. The uniformitarian approach is the basis of scientific thought. It states that events in the past, the present and in the future will always occur in the same way because their underlying principles are the same.

## PHYSICAL DEFINITIONS OF ANIMALS AND MAMMALS

Humans are animals, more specifically, mammals. Animals are organisms that move freely in space as opposed to plants that are firmly planted in their substrate. These initial definitions were proven to be only approximative, since there are some animals that are planted (*e.g.* sponges, corals, sea anemones), but all animals have motility - the ability to move quickly at will at least parts of their bodies –

Maciej Henneberg & Arthur Saniotis

tentacles, trunks, antennae, by means of contractile tissue such as muscles, while plants either do not move their bodies or move them very slowly by means of changes in tension in their tissues – flower petals closing and opening, leaves of insect-catching plants folding around their prey. Most plants produce organic substances by means of their synthesis from inorganic compounds, most commonly through photosynthesis, while animals obtain their organic substances from ingestion of other organisms.

Mammals (from Latin *mamma* – the breast) are animals feeding their young by secretion of glands in the skin of females. There is no exception to this rule. Even the most "primitive", egg-laying mammals – monotremes, platypi and echidnae – produce milk in the skin of the female ventral trunk. In the vast majority of mammals, the milk-secreting glands are organized into breasts – organs containing a series of milk-secreting ducts opening together on the elevated nipple. The milk-secreting ducts are transformed sweat glands. It follows that mammals have sweat glands. In terms of their evolutionary origin, mammals seem to be the organisms that thermoregulate the best. Next to birds, mammals rely on precise control of the energetic functioning of their bodies by maintaining nearly constant temperature in which all chemical processes in their organisms occur (homoeothermic). To do this, they must be independent of external temperatures in which their bodies function, as long as there is enough fuel supplied to their cells. This independence relies on the ability to conserve heat in colder external temperatures, to dissipate excess energy from the body quickly, and to cool the body in high external temperatures. These functions are performed by adipose fat tissue under the skin, hair, each strand of which can be moved in order to change the amount of insulating air trapped in the fur coat and to regulate air flow to the skin surface. The skin surface can be coated in fluid, the sweat, which cools the skin effectively by evaporation. These insulating and cooling mechanisms work well in a large variety of environments – terrestrial, subterranean, aquatic, arboreal and airborne – in contrast to birds who, though covered by fur-like feathers, cannot sweat and rely on exposure of their dry skin to air flow that is fairly effective because of their relatively large skin surface area in relation to their body mass. Small bodies have greater surface-to-volume ratios than large bodies.

With their stable body temperatures and fast-operating energy flow through their bodies, mammals are capable of swift and complex actions. These require coordination with sensory input. Control of complex actions and their coordination with sensory input, so as to orientate the organism optimally in its environment, is performed by nervous systems with highly centralized interconnections performed by the spinal cord and brain.

Various mammals adapted to various lifestyles in a variety of environments. Some of those adaptations were achieved by anatomical means – changing body parts so as to adapt them well to a particular function – *e.g.* a single-toed foot of a horse which has an excellent, energy-efficient means of propelling a large body on the ground, or a wing of a bat that allows it to fly swiftly and precisely. Alas, such adaptations limit a range of situations in which body can be used. A horse's forelimb is useless if there is a need to grasp a tree branch in order to climb up onto it; a bat's wing is useless when there is a need to run. As one of the oldest orders of mammals, primates, that include monkeys and humans, evolved by adapting their behaviours rather than anatomy of their bodies. Thus, they retained primitive anatomical structure while expanding their nervous system functions, especially the brain.

## SCHOOLS OF THOUGHT

People often report seeing things that actually are not there. For instance, at night in the backyard it may seem there is somebody standing in the dark near the shed: turn the light on, and it is simply a bush: shadows on an unfamiliar ceiling take on various forms; walking through the bush we see things behind every other rock; when somebody shows us an unfamiliar object we liken it to something already known to us; we can see things in ever-changing cloud shapes! Humans are beings who try to make sense out of what they encounter, based on what they know and what they have previously experienced. Science is a logical attempt to do just that. It seeks to enrich our experience, but it can do it only by making sense of observations.

During the 19th century, with the development of improved optical instruments, scientists began to describe geographical details on the surface of the moon and planets of the solar system. Mountain ranges were clearly visible on some, and on the moon many had characteristics like the craters of Earth's volcanoes - so they were called craters. Between mountain ranges were plains strewn with boulders and dark, flat-looking expanses. These were called "seas" and given poetic names like the "Mare Serenitatis" (Sea of Serenity). We know today that this sea of serenity is neither covered with water, nor more serene than any other flat, dusty, rocky and extremely cold piece of the Moon's surface.

The Italian astronomer Schiaparelli produced and published detailed maps of 'canals' on Mars. These showed a network of absolutely straight lines criss-crossing the surface of the Red Planet. Schiaparelli's work was reprinted and quoted by many astronomers [56]. Canals were an obvious sign of intelligent life, so stories about little green people living on Mars abounded.

Only several decades ago it turned out the 'canals' were an optical illusion created by old optical instruments, and yet NASA's discovery of fossilised life on Mars shows there was a grain of truth in the story. This grain, however, had nothing to do with the pictures of Mars that 19th century astronomers produced.

It was in the minds of Schiaparelli and his colleagues that intelligent creatures could live on other planets. How this belief became established, we will never know for sure, but one can point towards beliefs common to all ancient cultures: stories about elves, fairies, and gods who live in the sky. Some ancients even swore that they had seen fairies or heard gods' voices and witnessed their miraculous machinations.

Science and imagination go well together. While teaching at Cambridge University, the Viennese philosopher Ludwig Wittgenstein [57] observed in the 1920s that there was no such thing as a scientific "fact". What actually happens when a scientist talks about a fact, is that a state of things is described in his language as he has perceived it. Language is based on concepts and cultural experience, so the same observation may attract different interpretations depending on the user's personal background.

English has only one word for the white frozen stuff that falls in delicate flakes from the sky during winter: we call it "snow". Besides building a snowman or practising some winter sports, we have little use for it. The Eskimos, whose lives are spent on, and in the snow, and who use it as a building material, have some 30 different words for it. For them, it is numerous things, depending on consistency, position and practical use.

When we encounter a particular object or observe a phenomenon, we need to form an explanation to be able to make decisions. What do I do in this situation? How can we control this? Can I eat it? Will it bite me? Does it burn?

Humans had for ages lived in a natural environment, full of rocks and plants and animals. They had to recognise a multitude of living things to avoid falling prey, to eat without fear of being poisoned, to find enough food, and to cook it.

Testing by trial and error was completely unproductive. Since each animal and vegetable was slightly different, the best way to cope with the variety was to put them into categories, to create different 'kinds' of animals or 'kinds' of plants. For example, there are two basic kinds of living things: those who move (are animated: animals) and those firmly planted in the ground (plants). Among animals are those that live in water (fish) and those that live on dry land. It is not all that simple, because some animals live on land and in water - a frog, for example. That's not a big problem, as we can create a third category of animals: amphibians, those who live in both kinds of environments.

Each human culture, however technologically unsophisticated it may be, has its own 'kinds' of animals and plants and its members adopt these categories in everyday activities. Things covered with feathers are 'birds' and usually produce sounds, sometimes pleasing ones. 'Fish' have fins and live in the sea. 'Trees" are big, living things set firmly in the ground.

There are some problems with using 'kinds' to include all living things: for instance, mushrooms are firmly planted, but are not green and do not thrive on sunlight or carbon dioxide. This presents but a minor difficulty: we create a separate category - 'fungus' - into which mushrooms fit nicely. Each culture has its own system depending on their experiences, everyday needs and general beliefs in how the world works.

The Ancient Greeks loved philosophy. It provided a way of understanding the world that could be used every day. Philosophical doctrines were also a matter of life and death. The Athenian philosopher Socrates was condemned to death for teaching how to question artificially created moral values. His execution was, perversely, a triumph of early science, as he was put to death by drinking a poison – *cicuta,* a hemlock extract from plants. The process of poison preparation required more knowledge than killing by hanging, or by a simple blow to the head.

One of Socrates's postgraduate students, Plato came up with the idea that the world in which we live is but an imperfect image of an ideal world in which all things are pure and excellently organised. Hence, the cat we see is just an incarnation of the ideal cat which exists in another world. Plato's model of the world of ideals and world of forms was later elucidated by Platonists such as Plotinus and Augustine. Interestingly, for Plotinus creation was based on a series of emanations, each less perfect than the former – a possible allusion to the law of entropy in the universe.

The task of the Platonist was to discover pure harmony, by studying imperfect copies of ideals. In this way, he worked out a system of democracy that we still use in the government of various countries, in various forms. Another Greek philosopher, Aristotle - one of whose students was Alexander the Great, conqueror of a large part of the Old World - was a very keen observer of nature. He came up with correct anatomical descriptions of many animals and plants, and attempted to arrange them in some semblance of order. While much of western science has been posited on Aristotelian thought, Aristotle's legacy also provides a way for understanding teleological processes in nature.

Detailed studies of nature and the concept of an ideal world survived the Roman Empire, the Middle Ages and only became questioned, albeit partly, during the Renaissance. In this period, at the end of the 15th century, Europeans began to colonise remote lands and came in contact with new, strange creatures and plants. There grew an acute need to describe precisely the variety of living forms and to put these 'kinds' into some sort of an order.

In 1735, a professor of natural science at the university of Lund in Sweden, Carl von Linnè, also known by his Latin name Linnaeus, came up with two ideas still in use in modern biology. The first was that each individual plant or animal belonged to a special type (species) of a particular kind (genus). What had been created at the beginning of the world was an ideal type of each kind of animal and plant. Since then, individuals reflecting these ideal types had reproduced without change so that each new generation was a true copy of the old type. These individuals, due to the vagaries of environmental influences throughout their lives, could differ a little from the true ideal, but the variations were irrelevant and were just temporary corruptions of the ideal. It was therefore deemed appropriate to name each plant or animal by the genus and the species to which it belonged - or rather of which it was an example: a bit like passing on the family name, and first name. Hence, the common dog came to be called *Canis familiaris,* meaning dog of a familiar type. There were other dogs, like wild ones called *Canis lupus* (wolf).

Linnaeus' second idea was that there was a God-given order in nature, and he set out to discover this order. It soon became obvious that some animals were simpler than others, while we humans were presumed to be the most complex and at the apex of creation. This exactly mirrored human society, where peasants were uneducated, unsophisticated and earned a meagre living through long hours of unskilled work with their hands; the more sophisticated and skilled middle class had easier lives, while the upper classes were the epitome of refinement. All of this was presided over by the primates of the churches and princes of the royal class - the most excellent of people.

Linnaeus formulated the Ladder of Creation: with slimy worms close to the bottom, with monkeys, apes and humans at the top: he called the latter "primates". He also reached the conclusion that the only creature possessing wisdom was man. Hence, using Latin, he labelled people *Homo sapiens*: *Homo* being man, *sapiens* meaning wise. Women were overlooked.

As in a human society, the community of living things was divided into descending levels: kingdoms, classes, orders, and families. Humans, together with dogs, bats, rabbits, cattle, elephants and lions were members of the class of mammals, but only monkeys and apes shared our membership of the order of primates. Within this order, however, even apes were not good enough for inclusion into our circle. Hence, we are members of the tribe *Hominini*, while apes belong to a separate tribe, in the same family of *Hominidae.*

To fine-tune the system, a category including several classes within a kingdom was added - a *phylum* meaning "branch" in Greek. Classes of mammals, reptiles, fish, amphibians and birds are all members of the phylum of chordates.

As more and more animals and plants became known, various additional levels of classification became necessary. Thus subphylum, infra-order and super-family were created. The entire universe of living things was organised like a feudal society.

Incredibly, the system created by Linnaeus in the 18th century is still accepted in modern biology.

In October 2004, Nature magazine published an article titled: "A new small-bodied hominin from the Late Pleistocene of Flores, Indonesia" by P. Brown, T. Sukitna, M.J Morwood, R.P. Soejono, Jatmiko, E. Wayhu Saptomo & Rokus Awe Due.

The article states the following on page 1055:

"Order Primates Linnaeus, 1758
Suborder Anthropoidea Mivart, 1864
Superfamily Hominoidea Gray, 1825
Family Hominidae Gray, 1825
Tribe Hominini Gray, 1825
[Genus] Homo Linnaeus, 1756
Homo floresiensis (new species) 2004."

The hierarchical feudal system of government demolished by the French Revolution and the American Revolution in the 18th century, has been perpetuated in science into the 21st century.

The tragedy of modern biology is that there is no workable new system by which to describe the variability of life on Earth: just the outdated analogy provided by a hierarchical order based on the notion of superiority of certain classes and families [58]. This lamentable situation did not occur for lack of attempts at improvement: it is a result of these efforts not being widely accepted.

The first serious attempt to criticise the static hierarchical view of the living world was made in the mid-19th century, by two naturalists collecting field specimens of living things for scientific descriptions: Charles Darwin and Alfred Russell Wallace.

Working independently, they noticed that individuals, assumed to be reflections of true types of respective species, varied in expression of various specific characters. They also noticed that the offspring of various parents had different opportunities to survive and produce their own descendants.

Darwin likened this process of differential reproduction of individuals with various traits, to the well-known practice of animal breeders who selected for mating only individuals with desired characteristics. For example, only the strongest and best-looking stallions were selected to breed with mares of a specific breed of horses, whereas stallions who did not meet breeders' criteria were not allowed to mate.

Thus, as Darwin observed, there is no real difference between a species and a variety or breed. Various breeds of animals and varieties of fruit trees can be created by breeders. The same process takes place in nature, and in time one variety may replace the other, or one type of animal change into another. A new species could develop from an old one.

If a species could arise by means of a natural process, so can – given enough time - a whole bunch of new species constituting a new genus, and - after even more time - a new family, then order, *etc.* could arise without any intervention from the ideal world.

All of us are interested in where we came from and how we got here, but differing explanations of evolution have often confused the issue. This has been largely due to an increased number of fossil finds and the advancement of different scientific theories about the number of human species, and to different scenarios of human interaction and development. It is understandable, that increased quantity of new observations requires time and effort to be analysed and conclusions to be drawn, but there is a lack of integrative approach.

## CONSERVATION OF DIVERSITY

Had the Dodo not become extinct, the Dodo of today would not be the same as the Dodo we know from old descriptions.

### Extinction of Species

Species become extinct either because their members die out or because they evolve into something else. The only reason we do not call birds dinosaurs is that dinosaurs have been discovered after everyone knew about birds. Dinosaurs were bird ancestors [59].

# CHAPTER 6

# Children & Teenagers

**Abstract:** Evolution of specifically human periods of life – extended childhood and adolescence is discussed in the context of the entire process of reproduction. The essential characteristic of life is transfer of information from one generation to the next. Material structure of living things undergoes constant flux – old molecules are shed and new molecules are incorporated into cells – but the arrangement of parts of organisms remains largely the same, they contain the same information. Information is defined here as a particular arrangement of things. Such arrangement may remain the same even when actual objects are exchanged. Since individuals are perishable, the only way to maintain continuity of life is by passing information about organism's structure to newly produced individuals – the offspring. This process of reproduction commonly involves sexual combination of genetic information. The evolution of human sexuality with its hormonal regulation is discussed and biological foundations of human love and other complex emotions are suggested.

**Keywords:** Adolescence, Amenorrhea, Childhood, Fecundity, Genetic information, Menarche, Ovulatory crypsis, Reproduction, Sexual behavior.

Unique among mammals, human young do not go from infancy into a juvenile period ending in sexual maturity [60]. They have a long period of slow growth inserted between infancy and juvenile years. This is referred to as childhood. One can liken it to a larval period in the life of some invertebrates, *e.g.* insects. During childhood humans are asexual, they depend for food and protection on their parents, their slow-growing bodies are of size and proportions different from adults and, most importantly, their brain physiology and anatomy is different from that of adults. Children have more brain cells than adults, but some connections between those brain cells are not well insulated by myelin that ensures fast transmission of nerve signals.

Maciej Henneberg & Arthur Saniotis

Chemical regulation of brain function is different in children – they have no large quantities of sex hormones interacting with hormones influencing brain functions and secretion of neurotransmitters – chemical substances mediating nerve-cell--o-nerve-cell connections. Thus, children's brains are more "plastic" -- open to learning, less stereotyped in their behavioural patterns, with more possible connections between various nerve cells. As their life progresses, some behaviours and brain functions are repeated and circuits of nerve cells serving those functions become established, while disused cells and connections not exercised dwindle away [61]. Children learn fast and they learn for life. Since with their small, weak bodies children are not a competitive threat for adults, they are given protection and can exercise considerable latitude in trying various behaviours some of which may be seen as "inappropriate" among adults.

It seems that this larval stage of fast learning of vast amounts of socially useful information evolved in order to enable functioning of complex human societies and acquisition of extensive linguistic and technological skills necessary for successful human survival. Childhood ends in a short juvenile period in which individuals, though still sexually immature, can cope on their own in terms of procuring food and survival in a society (*e.g.* "street urchins" in poorer societies). Juveniles in modern societies enter a stormy period of adolescence during which sexual maturation occurs. Significant changes in hormone secretion appear and deregulate hitherto stable function and growth of an organism.

In the present-day individuals, the most striking feature of adolescence is the 'growth spurt' doubling childhood rate of growth, so that maturing girls and boys can increase their stature by 100 to 150 mm in a single year which means growing by 10% of their hitherto attained body height. In girls, the spurt happens usually a year or two before their first menstruation (menarche), in boys at a corresponding stage of sexual maturation, though this is less visible because production of sperm and ejaculate does not have a single-event nature. Though spectacular, growth spurt is less important, or as we will discuss below, even abnormal feature of adolescence. The most important feature of adolescence is activation of gonads (ovaries and testes) to produce fertile ova and sperm. This is a result of secretion of sex-specific hormones – estrogens and androgens -- in adult quantities. Girls mature approximately two years earlier than boys. Initial small changes in hormonal regulation lead to increased accumulation of subcutaneous fat. This accumulation eventually triggers off greater secretion of female hormones that activate the first ovulatory cycle. In most cases, due to social mores, there is no chance for sperm introduced during coitus to fertilise the first ovum a girl produces. Thus, the menstrual cycle is completed and bleeding occurs. This first menstrual bleeding has a Greek-derived name "menarche". The first ovulatory cycles of teenage girls are often incomplete, irregular. Thus, fertility of teenagers is lower than that of fully mature women. Demographers describe this as "teenage sterility". Despite this statistical observation, sensationalized reports of teenage pregnancies are commonplace. True enough, even the first ovulatory cycle can be sometimes fertile. The same is true of some other cycles of teenage women. It is simply a question of quantities – less teenage women's cycles produce fully fecund ova.

The moment of first ovulation, as marked by menarche, is considered an important event in the life of a girl. In some cultures, it requires a special celebration (*e.g.* in India), since it is interpreted to mean that a girl is becoming a woman. In ancient Rome, some girls aged 12 years having reached menarche, could be married. Ancient Indian writers encouraged girls to be married soon after menarche in order not to "waste their reproductive potential" [62].

Depending on the quality of their lifestyle, and genetic regulation of their physical development, girls can reach menarche as early as 9 years or as late as 17 years. This range is considered normal, though the majority of girls will start menstruating somewhere between 11 and 15 years of age. In well-to-do societies, the average age at menarche is about 12.5 to 13.0 years, in groups living in poor conditions it is delayed – 14.5 to 16.5 years. As we have already said, the moment of menarche does not mark full reproductive maturity of a woman. To achieve this full maturity, cycles must stabilize and the body will have to complete its physical growth. Hormonal changes at sexual maturation, when completed, slow down and then halt completely growth of bones in their length, thus fixing stature. The halt occurs about two years after menarche. Various tissues of the body, especially fat and muscle, will continue to grow for a number of years.

The highest realized fertility of women occurs during their twenties. This is due to biological fecundity of their bodies and lifestyle cohabitation. Young couples, cohabiting and fairly satisfied with their living arrangements, will copulate on average three times a week (although the variation is large: from less than once a week to more than once a day) [63]. This provides the optimal sperm count, and thus, the probability of conception. Too frequent intercourse lowers sperm count because testes cannot keep up their production with demand for the ejaculate. Trying to conceive by copulating too often is counterproductive. Once every other day seems to be the best (for conception – no reflection on other things). Even in happily cohabiting couples who have sex on average every other day, it takes on average three ovulatory cycles (approx. 3 months) to conceive. Human fecundability is low [64]. This may be a result of sex being used for social bonding, resources procurement or recreation.

Menstruation is the uncomfortable by-product of human love of sex for pleasure, social status and, sometimes, material gain, and our low fecundability. In all higher mammals, when a female ovulates, her womb becomes prepared to receive fertilized ova that will attach their placentae to womb's walls and then grow through a fairly long pregnancy. If, for whatever reasons, no developing embryos enter the womb after ovulation, tissues of its walls that were thickened and re-structured to support pregnancy, will start dwindling away and eventually detach from the womb in an event reminiscent of miscarriage – shedding of womb tissue supporting a failed pregnancy – and exit through vagina. This event produces period bleeding (menstruation) in females. Menstruation is rare in non-human mammals, because females become pregnant practically at each ovulation. Ovulating females signal their status to the outside world by changes in external anatomy of their genitals, smell and behaviour soliciting sexual intercourse. Thus, they can achieve pregnancy at each ovulation.

The hidden ovulation of human females (ovulatory crypsis) and the social and economic restrictions on sexual encounters, make it often unlikely for copulation to produce conception at an ovulation. In this way, the shedding of womb lining and menstrual bleeding become common. Yet, for most of human history, menstruation was not even nearly as common as it is today. The reason was lack of purposeful negative birth control (birth avoidance) and coordination of social and biological maturity resulting in much earlier age at marriage or beginning of sexual activity. Imagine, common in the past, situation. A girl gets married around her menarcheal age, becomes pregnant after a few months, goes through pregnancy, and gives birth. Then she recovers from her pregnancy (postpartum period), experiences lactational amenorrhea (women without sufficient fat storage will not ovulate while breastfeeding, and if their nutrition is stressed by lactation, they cannot accumulate enough fat to start ovulating for a long time while they breastfeed) and eventually weans her baby at 2-4 years of age. Only then she regains the nutritional status good enough for becoming pregnant again so she starts ovulating and, since she is cohabiting with a fecund male, falls pregnant again at about the third ovulation. Lets do the count. The girl married two months after her menarche. Thus she had 2 menses. She fell pregnant after 3 months of marriage, so she had three more menses. Then throughout her pregnancy (9 months), postpartum period (some 6 months) and lactational amenorrhea (18 months) she did not menstruate. Eventually, the baby was weaned and periods started again. Cohabiting, the woman took 3 months to get pregnant. Now, after 41 months since her menarche, she had experienced 5+3=8 menses. That makes it about 20% of her adult female lifetime, or experiencing menstruation on average in one out of 5 months of her adult life. It is likely that this pattern would continue throughout her fertile years. No wonder that in many cultures menstrual taboo has existed for thousands of years. Experiencing a period was a rare event in the life of a woman throughout most of history. Malthusian birth control changed it all. Nowadays, teenage girls and adult women menstruate for 90+% of their fertile life span. A very abnormal situation made easier only by modern technology that allows to hide the fact.

Boys undergo sexual maturation on average about two years later than girls. During adolescence secretion of testosterone surges and this produces a number of physical changes in boy's body, that, although not as clear-cut as in female menarche, allow to pinpoint the age of sexual maturation, independent of the growth spurt. Testosterone, besides promoting growth in size of testes and penis, seriously increases muscle mass and hairiness of the body, especially on the trunk and the face. Boy's pubic hair is usually not limited to the mons pubis, like in a girl, but extends up towards the navel, and onto the thighs. The chest hair may be abundant and in some regional populations (Aboriginal Australians, Europeans,

Ainu in Japan) facial hair becomes so abundant that without cosmetic adjustments (shaving, trimming) it may interfere with food intake and other daily activities. Cartilages of the larynx (voice-box) respond with serious enlargement producing very noticeable change in the length of vocal cords that deepens sounds emitted (breaking of the voice).

In both sexes, the change of the hormonal balance of the body resulting from the increased production of sex hormones, influences the functioning of the central nervous system through numerous neuro-hormonal feedbacks. Boys and girls change the way they pay attention to various elements of the world, especially human relations; girls become more sensitive, boys more aggressive. Like with any biological traits, there is a significant variation in behavioural traits of various individuals, and there is a wide overlap between boys' and girls' behaviours. Depending on their family and social environments, and, according to some researchers, on their genetic endowment, or epigenetic regulation, some girls may behave like boys and some boys like girls [65]. This includes their interest in the "opposite sex", thus, leading to a preference for individuals who are behaviourally, though not anatomically, their opposites. Girls in those circumstances may prefer girls instead of boys and *vice versa*. Previously considered "abnormal" this is quite a normal result of the overlap of variation ranges of human behaviours, neurohormonal regulation of behaviours and socio-cultural environment. We first of all are human, and only then female or male. Therefore, human emotions may be directed towards individuals who are not necessarily of opposite sex. In humans sex (biological, anatomical characteristics) may not be the same as gender (a feminine or masculine role with regard to other individuals). Since sexual encounters in humans became to a large extent decoupled from the simple biological act of reproduction, same-sex sexual relations may serve useful social and psychological ends in a particular set of circumstances. They will not produce biological offspring but may contribute to the continuity of human life in numerous ways by promoting productive actions or stabilizing social structure instead of aggressive disruption.

Since in humans control of sexual behaviour is largely conscious, occurring in the cerebral cortex, hormones and autonomic nervous system remain in the background. A complex of emotions related to love influences various intellectual functions extending far beyond interpersonal attraction and directly resulting interactions. Love-like emotions may provide motivation for artistic productions, commitment to a particular constructive course of actions, self-sacrifice, but also antagonisms and even hate. Together with predominant characteristics of the human brain – tendency to imitate, love makes our behaviour specifically human. Our emotions guide our actions while our actions are a series of imitations of earlier behaviours, those of others or our own. Since each situation we encounter differs at least slightly from previous situations, it would be advisable to alter our actions accordingly. This, however, requires innovative thought – a mental combination of imitative behavioural sequences that has not occurred before. It sometimes happens, but it is rare. More commonly we respond to new situations in stereotypical ways not precisely suited to exact features of the situation. If mismatches are minor, they may not even be noticed. Larger mismatches result in "misunderstanding the situation" or in not being able to predict consequences of an automatic imitative action. Humans are notoriously bad at planning or responding to rarely occurring challenges.

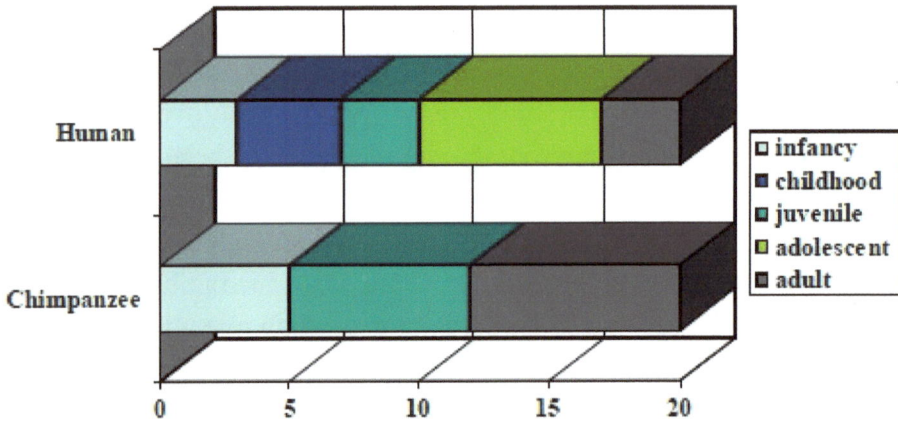

**Fig. (2).** Early part of life history. After Bogin (2001) [60] modified. Numbers are years after birth.

## LIFE IS INFORMATION

Most of us when walking through the forest will have no doubts as to which things we see are inanimate and which are alive. Rocks, soil, clouds, boots on our feet are inanimate while trees, shrubs, flies, grasshoppers and dogs are living things. How do we know it? Well, simply everyone in their right mind would see the difference between a tree and a cloud. The first is firmly fixed to the ground and does not change its shape from minute to minute. The second floats in the air nebulously as we watch. Wait! What am I talking about? It just does not make sense. We just called non-live things "inanimate objects" meaning that they do not move. Most of us faced with a task of coming up with a general rule of how to distinguish living things from other objects would either grope for bits of knowledge absorbed at school, give up, or simply think that it is ridiculous to discuss such things. "We all know" that plants, animals, bacteria, viruses are living things.

Lets try from the other end - one minute my mother talking to me from her hospital bed was alive, next minute she was dead. I screamed to the nurse for help, doctors came and after a while my mother was back with us. Two nights later, however, she died for good. What happens when a person dies? What does it mean to be alive? Is someone whose heart stopped beating dead? No, not really because resuscitation can sometimes bring them back to life. At what stage in the resuscitation process can paramedics give up and thus declare the person dead? You will probably tell me that to be alive means to be able to move, talk, think, make decisions *etc.* Yes, I agree. Is a totally paralysed person (thus unable to move and speak) not alive? These days they can, using a computer and a little machine called "voice synthesiser", speak to us although his mouth and voice box do not produce sounds. Maybe then the most important thing making a person alive is the ability to think? What then about people who are unconscious? Are they dead? Doubts abound to such an extent that scientists and medical experts cannot agree - thus cannot be trusted - and families of disabled or injured people turn to courts for decisions to declare their unconscious relatives dead. In this case "legally" dead. Is being legally dead less important than being really dead? What does it really mean to be alive? It seems that life is an entity occurring only as a combination of phenomena, an "emergent entity".

It is equally easy, or equally difficult, to define when human life begins. Some people say "at conception" and thus oppose abortion. Some others will say "at birth". In the recent past, there were some societies who did not recognise a new-born baby as fully human until a special act of acceptance was performed. Thus, if they did not like the neonate, for whatever reason, they killed it without hesitation nor remorse. Were these people inhumane? No, they would not dare to kill a human being. A human being, however, was only a person who passed an appropriate ritual. In the Christian culture there is a remnant of such ritual - babies must be baptised. Why? What tangible attributes of a child change at baptism? Why do some cultures perform circumcision? Why aliens who passed through the ceremony of naturalisation change into citizens? What happens? We create answers in a form of cultural categories. They help us to handle the variety of living forms and to cope with products of their incessant reproduction, but do not bring us closer to understanding the nature of life. We can buy a puppy, decide to slaughter an ox, or to give our kidney as a transplant to a sick sister, but many of us will doubt whether it is ethical to buy and sell living things, kill animals for food or to intervene into the natural course of somebody's life by replacing her body parts, or body fluids (*e.g.* blood), with "spares".

Finally, how do we define a human being? As an organism with two hands, two legs, *etc.*? What about war veterans whose limbs were amputated? In the 1960s, due to a pharmaceutical mistake a sedative medicine Thalidomide caused babies to be born without arms. Were they not human? When, as we often do, we discuss things on the internet with people whom we have never met, we form opinions of their personalities. Ultimately we treat our correspondents as real people. Sometimes we form friendships that are no less complex than those with people in our own town whom we meet in person regularly. It may well be that lurking somewhere in the web of internet there is an artificial programme able to converse with us as if it were another human being. Is such an electronic image a real person? When we "talk" to her on the internet she is.

A living organism has complex structure ultimately consisting of molecules built of atoms that in turn are built of elementary particles. It is debated whether these "particles" are physical real bodies or fields of energy and waves. The deeper we dig into the detail of life, the more it disappears. Some molecules in our bodies consist of many atoms and are quite complex, some others, like the molecule of kitchen salt - NaCl - are simple. All of them can be exchanged with the environment. The body gets rid of some salt through sweating and absorbs new salt molecules from the food. The same pertains to basic building blocks of muscles and brain tissue – proteins and fats. Depending upon an organ, over a period of days or years all its original parts are discarded and thrown out of the body while completely new molecules and atoms are absorbed from the environment. For example, the red cells in our blood are exchanged every 40 days or so. The blood that now flows in my arteries is different from the blood that was there in 1994. And yet it is unmistakably my blood. Its cells have my genetic signature. Any medical lab typing my blood will swear that it is the same as the one they typed in 1994. It is a bit like somebody who bought a Mercedes some time ago and still likes it very much. Tyres had to be replaced after a year, wheel rims, following some minor accidents, were replaced three years later, so were bumpers and the trunk lid. Two years later the frontal collision damaged headlights and the hood. These were replaced. Finally the engine gave up and had to be replaced. Transmission went the following year. Upholstery became so worn out that all seats were reupholstered. A better looking steering wheel was put in and the radio with MP3 capability. Finally the only original part was a cigarette lighter. Yet the proud owner of 30 years old Mercedes still maintains that he owns a wonderful car - "It was well worth buying a Mercedes - I still drive it after 30 years" he says.

Now we can try to formulate a definition of a living organism. Living organism is an object that maintains its structure and functions despite constant exchange of its substance with the surrounding environment. Sadly, as we know from our experience, it does not maintain itself indefinitely. Ultimately it dies. Many people will say, however, that it does not die entirely. It continues in the bodies of its offspring. Fair enough. But the offspring will die, too. How then is life continuing? It does so because many individuals exist at the same time and they produce new individuals to replace those who die. This constant process of reproduction is a bit like incessant replacement of molecules in the same organism. The difference is that here we deal with a set of organisms that produce offspring and die. The offspring will produce their own children and die. And so on and so on. Biologists call a group of individuals capable of sustaining reproduction over a number of generations a population. A population, unlike an individual organism, does not have to die, it can reproduce itself indefinitely.

During the process of reproduction parents transmit to offspring essential information about how to build an organism. This information is packaged as a genetic code - repeated sequences of four kinds of nucleotides, building blocks of nucleic acids. Each such block is a complex molecule produced by joining of the phosphoric acid with a specific sugar (deoxyribose) and organic base. The spine of each nucleotide is the same - phosphoric acid and a sugar, while bases differ between nucleotides. In the most common carrier of genetic information - DNA (Deoxyribonucleic acid) there are two strands of nucleotides containing four bases: adenine, tymine, cytosine and guanine. They are like letters ATCG of the code. The ribonucleic acid (RNA) in most organisms is produced on the template of the DNA as a mirror image of its single strand. Thus RNA is a single strand of nucleotides on the backbone of ribose and acid. Three of the nucleotides - ACG are the same, uracil (U) is inserted instead of tymine. The mirroring is done under the rule of complementarity so that nucleotide A on the DNA always puts nucleotide U into an RNA strand, while C always couples with G. After this transcription of the code from DNA into RNA the single strand of RNA is used to direct production of proteins - basic blocks of our body. Proteins consist of amino acids, small organic molecules that are strung together according to the RNA nucleotide code. Since there are over twenty amino acids commonly encountered in our bodies their sequence cannot be coded simply by single four letters of the genetic code. Each amino acid is described by three letters of nucleotide code. For example the amino acid alanine is coded for by a sequence GCA in the RNA corresponding to sequence CGT on the DNA code. Since there are 64 possible three-nucleotide combinations and only about 20 amino acids to code for, some amino acids are coded for by several sequences. Aforementioned alanine is also coded for by GCU, GCC and GCG triplets. Note that the first two letters of all four codes are the same. In some viruses DNA is not present and their codes are reproduced from RNA. Not all parts of a very long DNA molecule consisting of thousands of nucleotides, contain informative sequences of nucleotides. Such coding sequences are separated by long stretches of informational garbage that, however, are structurally useful. One can liken DNA code to a necklace consisting of beads separated by long stretches of bare cord. Only the beads contain useful information, but the cord itself is needed to hold the entire necklace together.

The code contained in the nucleotide sequences is universal for all living things. It works the same in a bacterium and in an elephant. What differs are the specific sequences of nucleotides that convey different messages: one compelling amino acids to build proteins characteristic for a bacterium, the other for various cells of a large and complex organism.

One can easily ask - why all this trouble - would it not be better to simply pass proteins from parents to children? It would, but for one thing - protein molecules are large and unstable. What happens when a protein molecule deteriorates is easy to smell - just leave a piece of meat outside the refrigerator for a few days. Proteins must be constantly re-built in the cells of the body. The machinery that does this is the DNA-RNA coding sequence. DNA molecules are very tough - their fragments can survive for hundreds of years. Yet they also break down from time to time. If such a breakdown occurs in the coding region it will affect the quality of the product - a wrong protein will be made and this may throw off the whole machinery of the workings of a cell - a tumour may start growing out of this affected cell, or if the cell is in a gland producing an important hormone, the levels of this hormone may fall. Such a breakdown affecting the quality of genetic code is called a mutation. Mutations are inevitable because DNA molecules, like all molecules in the world, must succumb to the rule of energy loss - they tend to go from more organised state that requires higher energy levels for its maintenance to a disorganised state that is energetically less expensive [66]. This rule is known as the second law of thermodynamics. DNA molecules must break down sooner or later. They are like a car (car is also subject to laws of thermodynamics) - no matter how good it is, it must break down sometime. The only way to keep the car running is to keep replacing the parts. The replacement of DNA parts is called replication and happens regularly in all cells. During this process the code from the old molecule serves to string together in the appropriate sequence nucleotides of the new molecule before the old molecule breaks down. This new molecule now will be used as a template to produce yet another one, well before it breaks down due to the passage of time.

The code in which Shakespeare wrote his plays - paper and ink -- is also subject to the laws of thermodynamics. For many plays the original paper and original ink have deteriorated and no longer exist. Yet we know each word the playwright wrote centuries ago because texts were copied from one edition of his works to another. Today we find Macbeth printed in synthetic ink on chemically processed paper, yet it is the same Macbeth that was written with a quill pen on rough wood pulp barely held together by natural glues. Actually no ink nor paper are necessary, we can read the text of a play on a computer screen.

Life, therefore, can be defined as the perpetuation of information about the structure of organisms. This perpetuation, however, unlike reprinting of works by famous authors, has no proof-reader. As mentioned before, replication of the nucleotide code is not perfect - long DNA or RNA molecules are subject to normal rules of physics and errors in the substitution of nucleotides occur. The information passed from parents to children is not perfect. It has a bit of error here and there. Like with a book, one misspelled word per chapter does not make much difference to making sense out of a text. When in a series of reprintings more and more errors creep in, the text may become unreadable. Then no one would buy the book and it will become extinct. Why then has life, based on fragile nucleotide codes, not disappeared?

Errors in the genetic code are checked by natural forces. Over time ways of minimising the effect of erroneous DNA replication have developed. The proof-reader checking the genetic code is not very meticulous. It will pass almost anything that works. If an error occurring in the genetic code is small enough not to interfere with the development of a workable organism, a child will grow up and in all probability produce its own offspring, though its bodily structure will not be exactly as prescribed by parental DNA.

Albinism provides a good example. Albinos have a genetic defect that prevents them from producing melanin, the common dark pigment of the skin, hair and irises [67]. Their skin is very fair, hair is almost white and their pale blue eyes have a red tinge since the blood flowing in tiny vessels of the iris can be seen through its unpigmented surface. Many albinos lead normal lives and start their own families, but their skin is very sensitive to sunlight and they often develop eye problems resulting from overexposure to ultraviolet radiation that is not screened by the pigment. With the avoidance of some sunlight and with a bit of medical intervention albinos can live normal lives. Imagine, however, being born an albino on African savannah several thousand years ago - in all likelihood you would not have survived to adulthood. Chances of having a spouse and producing babies would have been seriously diminished, especially in comparison to those people whose skin and eyes were well protected by lots of dark melanin.

The natural forces checking the "correctness" of the genetic code are called natural selection. Their operation is clumsy and often not immediately visible. Natural selection will easily eliminate obvious genetic defects that prevent an embryo from developing normally. One of such defects is called anencephaly. This Greek term means lack of brain. An individual with this defect will fail to develop the entire central nervous system - no brain and spinal cord are formed in an otherwise normally growing embryo and foetus. Anencephalic children die before they are born and in earlier times were miscarried.

Albinism is not dealt with by natural selection with the same severity as anencephaly [68]. Operation of natural selection on other human characteristics is even less obvious. It may consist of just slightly altering chances of dying or of having progeny. Its operation may also depend on circumstances in which people live.

Colour blindness is a genetic condition [69]. It affects males more often than females and to a varying degree. Typically, colour blind people will have difficulty distinguishing between green and red colours. In most circumstances this is not a big deal. After all, we still enjoy watching black and white movies and the whole generation grew up on monochrome television in 1950-s. What has the ability to distinguish between green and red to do with making babies? Precious little. A colour blind person will also rarely cause a traffic accident by mixing stop and go signals since everywhere the red light is on top and the green at the bottom of the signal. Consider, however, surviving by hunting and gathering of wild foods in a complex environment when most of your food has camouflage coloration that requires good colour vision to be noticed. Quite an impediment.

Natural selection can slowly chip away at minor impediments to the way an organism works. The only rule here is that it will eliminate anything that interferes with efficient reproduction – survival of parents and production of viable offspring. Unlike the human proof-reader, natural selection has no set goal. It simply happens because the genetic code can replicate itself only through building of organisms that will grow and function efficiently in a given environment and be able to pass the code to their offspring. For this reason some happenstance errors in the genetic code, instead of being eliminated by selection, are favoured by it if they happen to be useful for survival and reproduction in an altered environment.

We have just said that in albinos a complete lack of melanin darkening the skin is bad. Quite true. Even a partial lack of skin pigment, like in many fair-skinned and blue-eyed people of European descent, can be dangerous if they live in climates with intense sunshine. Incidence of skin cancers among "white" Australians is a serious problem as their skin has too little melanin to protect it against massive penetration of UV rays at lower latitudes where Australia is located. Light-skinned people visiting Australia are advised to cover themselves with clothing and wear hats or to use various sun-blocking lotions. A nice tan is no longer a sign of health and vitality. Why then are many Europeans fair-skinned? Human biologists hypothesise that when people first entered Europe they encountered a cold climate with commonly cloudy skies [70]. Under such circumstances melanin was not much needed. To the contrary, its presence blocked small amounts of UV radiation available in such a climate from entering the skin. It so happens that some sunshine must penetrate the skin to allow the body to produce vitamin D. Under the cloudy skies of Western and Northern Europe those individuals who happened, due to a chance genetic defect, to have poor production of melanin, became healthier as they could use poor natural light to produce necessary vitamin. Hence, instead of dying, such de-pigmented people fared better than their normally coloured relatives and produced more offspring over many generations. Today, humans have this genetic anomaly of fair skin--blue eyes--and blonde hair occurring so commonly that it is no longer considered abnormal. In some circles it is even considered attractive.

Although genetic errors - mutations - may sometimes be useful and beneficial, most are deleterious. The chemical quality of DNA and RNA is such that occurrence of errors cannot be prevented. Mechanisms minimising bad effects of mutations do not interfere with the chemical structure of polynucleotides. They are rather based on a principle of cover-up. Quite different from your car that has only one engine, the genetic code, to prevent breakdowns intervening with the normal operation of the organism, simply puts several copies of the same code into one body. As if a car had two, or even three, engines. When one breaks down the other will keep powering the vehicle. A rather clumsy way of doing things, but it works. Since in nature there are no intelligent designers and proof-readers, whatever works, however clumsily it is put together, goes. At least initially - streamlining of operations takes time, but it is eventually executed by natural selection. Natural selection ultimately favours those organisms that can achieve reproductive success using as little energy as possible.

There are two basic ways of putting several copies of the same code into an organism. The first consists of stringing into the long sequence of polynucleotides same segments several times, that is, like putting the same word into the sentence several times. Such words are called repeats and occur commonly in DNA. When one repeat breaks down others will still provide correct information for the operation of chemical mechanisms in the cells.

The other way is to multiply the entire genetic material. Most organisms have double sets of DNA - they are called diploids. There are, however, those who have triple, quadruple or more sets of genes. Some common varieties of cultivated plants, such as wheat, are of this kind.

Sex has been a great invention of nature designed to cover up poor quality of the genetic code. Since the code degenerates as time goes by and errors in nucleotide substitution accumulate, the way to cover up some of these errors is to combine genetic endowment of one individual with that of the other. There is a chance that where one copy of a code has erroneous information the other one will have the correct copy and *vice versa*. Had the entire sets of genes been combined every generation pretty soon the amount of DNA in the cell's nucleus would become enormous and unmanageable. The cell would become like some university libraries. They are so keen on collecting and holding printed information that they become the biggest buildings on campus, but a large library does not make a university the best school in the nation. What really counts is how academics and students use information.

The idea of sex is to combine half of the genetic code of one parent with the half of another one. Sexually reproducing organisms have at least two copies of the code for each characteristic - one obtained from the mother and one from the father. Such set of genes we call diploid. Paternal and maternal code for a given trait, say, blood group, may not be identical. For example the father's code may lead to blood group O while the maternal one codes for A. We call such various forms of the code for the same trait alleles from Greek "allos" meaning different. Every time the germ cells - ova and sperm - are produced the genetic endowment is split in half and only one allele for each trait gets to the germ cell. Thus a sperm or an egg will have only half of the diploid set of genetic information. We call such halved set haploid from Greek "haplos" for single. Since mother and father each produce a lot of ova and sperm during their lifetime the advantage of splitting the code is to produce some germ cells with one variant and some with another variant of the complete parental code. If a parent had one defective allele and one well-functioning one, every other ovum will carry only the healthy variant of information. When, following the interesting part of sex, sperm and ovum finally come together they combine their respective halves of genetic information. The first cell of the future baby will have once again the diploid number of alleles - two for each trait, but their combination may be nothing like those seen in father and in the mother. If the combination for a given allele happens to be that of a defective allele from the mother and the defective allele from the father the child will have the defect in its body. Depending on the kind of defect such poor child will either be unable to develop at all, or will develop poorly and die early before having any chance to pass its defective genetic code to the offspring. Eventually it may develop well enough to have a go at making babies, but with its physical defect will not be very effective in this task. In short, such an individual will be subjected to negative natural selection as unfit to live or to reproduce effectively. If, however, a defective allele of the father combines with a healthy allele of the mother, or *vice versa*, the child will be fine. It will obviously be fine also when a healthy-healthy combination happens.

Sexual reproduction allows carriers of various defective alleles to live normally and to reproduce successfully although half of their sperm or eggs containing wrong information will be wasted. In the meantime environmental conditions may change and the previously defective allele will become useful, like in an example of poor melanin production.

To sum up - life is perpetuated by continuous copying of genetic information from one polynucleotide molecule to another. Although each single molecule of DNA or RNA is chemically unstable, the information may be kept almost indefinitely by its replication and trimming out of errors by natural selection. How long is "almost indefinitely"? In some instances the same information is maintained over scores of millions of years. Fossils found by palaeontologists indicate that the information about how to produce a bird's feather has been around for nearly 100 million years. Have a look at a feather under a strong magnifying glass - its structure is quite complex, and so it was tens of millions of years ago. Bones, teeth and skulls of people who died 30 thousand years ago look very much the same as our own. Since the average age at which people have children is, approximately, 25-30 years, this means that the genetic information on, say, how to make a molar tooth has been passed from parents to children >1000 times and yet remained unchanged. Take a mirror and look at your tooth. It contains information about your ancestors whose bodies have turned into dust a long time ago.

Although the stability of the genetic code in the face of constant exchange of nutrients with the environment and the incessant production of offspring is amazing, errors in the genetic code are quite common. For example one in ten thousand babies is born with a defective gene for growth of bones in length. This condition is known as achondroplasia [71] because cartilage responsible for bone growth is defective. Individuals affected are very short and often are called dwarfs.

The same condition in dogs has been considered to produce a cute effect of a "sausage dog". For this reason breeders selected such dwarfed individuals and by mating them with each other produced well-known and affable breed of dachshunds.

One in ten thousand is by no means a rare condition considering that every year some 50 million children are born world-wide. Five thousand of them will have the defect. If you live in a rather small city of 100 thousand people, almost every month an achondroplastic baby will be born in your local hospital. Table **2** gives the frequency of some better known genetic errors. Since these errors change the gene they are called "mutations" after the Latin word for change.

In the face of the constant onslaught of mutations it is only the abundant copying of the genetic code and relentless trimming by natural selection that keep the information same over millions of years. What seems like a stagnant situation of boring copying from generation to generation of the same information is in fact a very dynamic process of copying and elimination of errors. Anything that will allow more effective copying and passing of undamaged information from generation to generation is favoured by natural selection because it improves reproduction.

One can ask why the whole business of life as described here has occurred at all and why is it so clumsy - long unwieldy and unstable molecules, mortal, ageing individuals, exciting but also unnerving process of making and raising babies. The answer to the first part of the question is - there was no particular need for life to arise. It did so because some four billion years ago a combination of chemical and physical conditions on Earth, in accordance with universal laws of physics and chemistry, happened to promote cyclic reactions between long molecules. This points to the answer to the second part of the question. Since it was just the coincidence of the combination of molecules and temperatures, the process is as clumsy as a lump of clay being slowly pushed down the hill by rivulets of rain water. Yet sometimes such a lump of clay, or an eroding rock, may assume interesting and pleasing shapes. So does life.

The process of cycling complex chemical reactions involving long molecules consisting of repetitive parts - like nucleotides in the DNA - once started kept on going by replication, development, production of offspring. Only those individual lines that reproduced themselves efficiently were able to continue. Errors, especially grave ones, led to inefficiency and disappearance.

We know these general things. Many of the details of how life started are still debated by scientists. In one way the process must have worked because we are here. In the other it did not work all that well because we have many inadequacies. For example, we do not have wheels although the wheel is much more efficient way of getting around than these jointed, complex legs we have. We must pay for food and cook it while plants get their nutrients from sunshine, carbon dioxide and water in the soil. No matter how heavily you breathe, how much water you drink and how much you expose yourself to sunlight you will die of starvation because at some stage in our evolutionary sequence we did not put the green chlorophyll into our cells. What a pity - we would be so much better off with beautiful green tinge of our skins soaking up nutrition from the sun free of charge while we read or work. Who is to blame for this omission?

Although we happened to miss the chlorophyll boat, we did quite well further down the line when we caught up with the idea of efficient passage of information from generation to generation. This basically required better efficiency in terms of energy consumption. As mentioned before, the basic criterion by which natural selection assesses fitness of individual lines of descent is the energetic efficiency with which they reproduce information from one generation to the next. If the only way to code the information and to pass it on is by replicating DNA then it takes a lot of metabolism to build up and maintain each cell, to develop the body and to produce babies. If, however, one can usefully code biologically meaningful information in other ways, one wins. These other ways are intellectual. Understanding of the environment allows us to avoid some disasters and to place ourselves in the most advantageous locations. It allows us to plan actions that will lead to catching food, finding shelter or protection of other human beings. We can even plan actions that will reshape our environment so as to produce shelter or add a covering of animal hide to our body. We can pick up a useless blunt stone, break it up into sharp pieces and kill an animal by cutting its throat with a sharp blade. Instead of waiting for the transfer of our genes to pass to young people what we know about getting food, or protection against inclement weather, we can simply call in youngsters and tell them how to do these things. Finally, in order to warm up our bodies we do not need to eat organic substances and put them through rather complex chains of metabolic processes producing body heat. We can simply gather a pile of wood and set it alight. We can do this only because we know all the steps that lead to production of useful warmth out of dead wood.

However, we have perfected the use of non-genetic information to protect and perpetuate our lives we are not alone in inventing various ways of doing this. For instance, many social insects build artificial shelters - anthills, hives, termite mounds. Bees who have found a rich source of nectar can give other bees detailed directions of how to get to this food by executing a complex dance at the entrance to the hive. Some crabs use empty shells, or even discarded cans, to cover their delicate tails. Chimpanzees in the wild will pick up sticks for use as tools or will use stone anvils and hammers to break nuts open.

We can manipulate the environment according to information we possess. This information enhances our survival with the minimum use of energy. The more energy efficient the passage of information becomes, the greater our ability to survive as a group. An ancient leader speaking in an assembly of his 500 people could inform all of them about new opportunities of growing food or building houses. If she wanted to reach 10 thousand people she had to sent messengers on foot or on horseback to other 20 communities. These messengers and their horses had to eat and sleep on the way, had to expend energy on getting people in those other communities to listen to them *etc*. Today, we just put the president on the radio and his message is conveyed immediately to millions of people as long as they can power their radios – a minimal use of electricity. Energy-efficient.

Since life is really a perpetuation of information, it is a regular outcome of evolution to produce organisms that will be more and more energy efficient in acquisition, maintenance and use of larger and larger amounts of information. The story of human evolution is just one example of this general rule. Evolution of social insects is another, although less expansive. With the recent possibility of the presence of life on other planets in other solar systems [72], we may expect this rule to be carried out to its logical conclusion. We can predict that we are not alone in the universe. Study of the story of human evolution is not just an exercise in learning a specific unique history of one creature, it is an attempt at understanding how intelligent beings can evolve anywhere where life can exist.

The story of how slimy little cells evolved to fly jet planes and play electronic games is one of the most exciting and the rarest in the universe. Going through big spectacular eruptions is a stuff of many stars. Crashing violently into the planets is a fate of comets. Spectacular but short-lived. History of tiny chemical sequences of nucleotides laboriously putting proteins together for billions of years and slowly building bigger bodies with brains in them is much less spectacular, but it is also much more exceptional, and, even more importantly, it is about us.

CHAPTER 7

# Similar but Different

**Abstract:** Individual biological variation and its sources are described. The role of genes and environmental factors in the formation of variation is explained, basic methods of measuring variation given and examples of variation presented. Distribution of human variation within and between populations is discussed with the conclusion that the concept of "race" is not applicable to humans. Differences between people from various geographic areas are small, not sufficient for meaningful separation of humanity into distinct subunits comparable with breeds of some animals. The process of microevolution as a continuous alteration of variation ranges is exemplified by recent changes in the prevalence of genetic defects, anatomical variations (*e.g. spina bifida occulta*), reduction of human brain size by about 10% in the last few thousand years, shortening of body height when hunters-gatherers adopted agriculture and subsequent increase in body height in post-industrialised world, and by alterations of head shape in some nations in the last few hundred years.

**Keywords:** Albinism, Microevolution, Nutritional status, Phenylketonuria, Race, Spina bifida occulta, Variance.

Cats are cats are cats. But each is a bit different. Some have darker coat, some lighter, some have three white feet and one black, others two black and two white. Some are fat, some thin. Well, cats are artificially bred and variously fed by their owners. Not a big wonder then that being of the same species (kind) they are each a bit different.

Take field mice or common fruit flies. They all look the same to an average person. Surely they must be the same as nobody breeds them for special characters and they all eat the same. Not true. When in 1920 American geneticists began to keep fruit flies in their labs, it turned out that some had brown eyes,

Maciej Henneberg & Arthur Saniotis

some red ones, some in addition to normal one pair of wings grew another pair of wings. Number of hairs growing on their legs differed too.

The same applies to field mice. Had you caught 100 of them you would see differences in size, coat colour, length of tail, *etc*. In fact, each individual animal or plant is unique. This is due to the fact that the only thing it has in common with its parents are sets of genetic information put into the first cell of the new organism. In case of sexually reproducing organisms, even the genetic information is not completely shared with each single parent. A mother provides half of the child's genes, and a father provides other half. Some genes exert specific influence on the body in the presence of particular genes, while another set of genes in the presence of other genes. Thus, the very combination of maternal and paternal genes may make a child different from each parent. Add to this the fact that genes are just carriers of information. This information must be read and translated into actual building blocks of bodies by several steps of chemical reactions. How these reactions occur depends on the environment in which the body grows – shortage of certain elements in food, lower or higher temperature, exposure to radiation, *etc*. all influence the final product, a grown up individual organism. Genes and their reading mechanisms also interact with each other inside the cells of developing body adding complexity to expression of the information contained in DNA inherited from parents [73].

The first cell of the new organism - a zygote - being given its set of genetic information must cope with the process of growing into a complete body on its own and then cope with all the vagaries of its adult life. True enough, both animal and plant parents often provide newly produced offspring with food and protection for the first stages of life. This may take form of food stored in a kernel of a seedling plant, hard shell surrounding an egg, or keeping the young inside mother's body like in sharks, some lizards and most mammals. Parental help sometimes goes further in a form of post-natal protection and teaching.

All these are given to an offspring on a "take it or leave it" basis. No parent can get inside a body of a child and fiddle with its metabolism. Thus, two growing organisms, even given exactly the same genes, will encounter slightly different conditions during their development and may end up being of different size or having different behaviours.

A popular saying that "it is all in the genes" is not true. Genetic endowment may be overridden by environmental manipulation. A simple genetic defect causes condition known as phenylketonuria (PKU) in people. It is quite common – depending on a country 1 in every 5000 to 100000 people is affected. The genetic defect consists of the incorrect protein metabolism that produces high levels of one of the amino acids - phenylalanine - in the blood. Although phenylalanine is a normal component of animal protein, in high concentrations it acts as a slow poison. Thus, a child with PKU grows slowly, is retarded in its intellectual development, and if not treated may die in the first few years of life. The treatment, consisting of special preparation of proteins consumed by a patient is now available [74]. It is rather costly. Yet, provided that a child is treated from early on in its infancy and maintains a special diet throughout life, it will grow into normal, strong, healthy and intelligent human being despite having a defective gene in every cell of its body.

Although it is not all in the genes, the other popular saying "you are what you eat" is not true too. Tall stature is commonly considered a sign of health and physical strength. In community health surveys, stature of individuals is considered to indicate their "nutritional status". Yet if a person has both parents who are short, no matter how well she is fed during childhood and adolescence she will not grow very tall. True enough, she may be a bit taller than her parents, but not much. In the studies comparing body height of adults from the same population some of whom come from very poor socio-economic conditions, some from very good conditions, the average difference in stature does not exceed 5%-6%, that is 0.1 m (100 mm) [75].

With the advent of fashion for thin, emaciated-looking bodies we all talk among our friends about how easy it is for some of the people to stay thin even without much dieting, while others grow fat no matter how much they try to limit their food intake and increase physical activity. Knees damaged by too much jogging, anorexic behaviours, psychological problems, and thousands of dollars left in coffers of various therapists and gym owners are a tangible measure of how difficult it is to artificially alter our body shape and its composition (fat/lean body mass). Even larger sums are earned by publishing books, magazines and articles about absolutely sure ways to lose weight and stay healthy, by running "health clinics" and selling the latest exercise gear. In fact, fast walking in your normal business suit and shoes or in an average dress and "sensible" shoes can lose you the same amount of calories as exercising in a special outfit and footgear costing you some $200 plus gym fees.

The truth, as is often the case, lies in between the two extremes. A lot is in the genes but you can modify it a bit during your life. From time immemorial humans tried to cope with a dilemma: we are all human but each of us looks different. Within your own family you may have a brother with black hair and a red-haired sister. Your father may be a very intellectual type while the uncle is an outdoor type. Many people in your own neighbourhood will look very familiar. Not only their body sizes, shapes and colours, although as variable as within your own family, will all fall into a familiar, "acceptable" range, but also their style of dress and demeanour will be familiar and comfortable.

Yes, the fashion goes beyond what we wear and how we behave. The very anatomical content of our most private possession - our body -- falls victim to fashion trends. Recently it is fashionable, especially for women, to have cadaveric look of emaciated victims of a famine, while males are the best when looking like brutes with bulging muscles. It was quite different in the Victorian era - women were expected to be well-rounded, while portly figure in males indicated their well-being. What the 21$^{st}$ century will bring no one can be sure.

More importantly, however, each culture on earth had to somehow conceptualise the fact that inhabitants of other countries or continents look different from ourselves. Some are dark skinned, others diminutive in stature and less hairy. Some have thick lips, some thin. Some people of other kinds look dignified, some funny.

Growing up in our family and our neighbourhood we become not only accustomed to certain varieties of face, nose and lip shapes, but also to certain subtle grimaces and eye movements. We can also easily pick up very subtle differences between individuals, even between look-alike siblings. A mother of identical twins will soon learn to distinguish which one of them is Mary and which Jill. When we go to foreign country the situation may easily be different. The first few days in Thailand or in Gabon everybody will look alike to us. Slowly we will be able to pick up different ranges and combinations of variability in facial details. If we stay long enough, differences between individuals will seem normal and very familiar.

In 1986, we have moved from Texas to Cape Town, South Africa with my wife Renata. The variety of people living in the city was astounding. Besides immigrants from Europe and their descendants there were native people of the region - Khoisan. Names Khoi and San replaced today somewhat derogatory terms of Hottentot and Bushman used by early European colonists, though now even those names are considered inappropriate by some. Khoisan people are the oldest inhabitants of Southern Africa, speak peculiar languages full of clicks produced by fast movements of the tongue against palate and teeth, and are basically cold-adapted. Winter nights in the vicinity of Cape Town are freezing with snow falling frequently in the nearby mountains and sometimes in the city. Thus, indigenous people have relatively light skin, and eyes protected by double fold of fat-filled skin on upper lids, just like people of Mongolia and adjacent regions of Asia. The hair is very tightly curled to the extent of being bunched up in small peppercorn-like dots on the scalp. Body size is rather small. Dutch colonists brought in the 18th century to Cape Town rebel leaders from their Far Eastern colonies (today's Indonesia). In the 19th century, the British dragged some Madagascar slaves and people from India as indentured workers. Considered "typical" black people of Africa speaking languages described as "Bantu", came into the present-day South Africa about 2000 years ago but settled in the North-Eastern part of the country. Only after the European colonisation some of them drifted into Cape Town area. Furthermore, since Cape Town was for three centuries an important seaport on routes between Asia, Europe and America many crew members from all over the world left their genes there.

Although initially the variability was bewildering, soon enough it became normal. The University of Cape Town and the Medical School, where we both worked, were strongly opposed to apartheid and promoted among the staff and students complete non-discrimination. Hence, our colleagues and students sampled the whole spectrum of human variability form the city and the country. In day-to day work skin colour and nose shape differences soon ceased to be noticed and individualities came to the fore. The workplace looked as normal as anywhere else in the world. To this day people of Indian and Khoisan origin remain our best friends on a par with some descendants of English and Dutch colonists.

One of the technicians in the Department of Anatomy, Neville Smidt, a very talented organiser of fieldwork and excellent pro-sector of anatomical specimens was small, slender, had very dark skin, broad nose and nearly black eyes. I am tall, heavily built, blue eyed and my hair was in my youth blonde. Yet, after a bit of discussion of our ancestry we found that we share German origins. His grandfather was a "pure" German, and so was mine! So much for "racial" differences. We see what we believe in. True enough, Neville's grandmother was Khoi while mine was Polish, but so what? When we moved in 1996 to Adelaide in Australia, walking down the main street I had an uneasy feeling of something being out of the ordinary. It took me several days to realise what it was – the vast majority of faces on the street were pink instead of a healthy, normal mixture of various shades of brown.

How do we describe this bewildering variability of sizes, shapes and colours? There are basically two ways of doing it. The first is creation of clearly separated categories such as "black" and "white" or Asian and European. Can you, by the way, tell where exactly is the border between the two continents? The other is laborious metric measurement of various bodily characteristics and statistical calculations of averages and ranges of variability. No normal human being can keep doing this while meeting friends or customers.

Thus the concept of "race" has been created. We all know that dogs, horses, cats *etc.* have different breeds. Why not people? We also know that dogs are not only bred for coat colour, muzzle length and ability to run but also for their "disposition", a suite of behaviours. There are fierce breeds and there are gentle, sleepy ones. Why not people? In this way misguided and oversimplified concepts have been created of African people being dark skinned, flat-nosed and phlegmatic in their behaviour, while Europeans are tall, blonde, blue eyed and intelligent.

We create mental templates in order to escape the bewildering variety of humans. These templates are called races. They have very tangible practical applications: Holocaust, apartheid, segregation under "equal but separate" legislation, even 1996 Olympic stables of highly trained black runners and white coaches. The only practical results are human death, suffering and exploitation.

Laborious measurement remains the only honest and objective solution. Although, at first glance impractical, it produces very practical results. They are as follows. When one takes such important, and often misused by racists, characteristics as human brain size, it will turn out that there are differences between average brain sizes of various regional populations. These, however are much smaller than differences between individuals within the same population. True enough, male brains are on the average about 10% larger than female brains, but this difference fades when two males are shown to have brain volumes differing by 50 or 60 percent. An average human brain has a volume of about 1350 ml (three pints), but there are many quite normal individuals whose brain volumes are below 1000 ml and equally many people with brains larger than 1700 ml who are not geniuses. Since the density of brain tissue is about 1.036, the volume nearly corresponds to weight - from less than one to nearly two kilograms [51].

It is difficult to show any practical correlation between overall human brain size and intelligence. Suffice it to say that women with their smaller on average brain size usually score the same or slightly better than males in various tests of intelligence and mental aptitude.

To sum up, about 50% of total variation of human brain size can be found among same sex individuals in the same population while sex differences are about 25% with only another 25% being due to what some people would describe as "race" [76, 77]. It would be repetitious and boring to repeat the same results obtained for body weight or for distribution of some blood groups. The fact remains that in most characteristics we differ more from our immediate neighbours than the average for our population differs from an average for another one.

Since each of us is unique it is no good to describe human populations or even the entire human species in terms of averages. A measure of variability doing justice to short and tall alike is much more appropriate. In statistics such measure is called variance. It is defined as an average squared difference between a value of a character in an individual and the average for the population. Why squared? Because squaring makes all differences positive and exaggerates rare large values while playing down most common small differences. For calculation of variance it does not matter whether your height is 10 cm below or above the average, what counts is that you are 10 cm different from the average.

Mathematical formula for calculation of variance is: $V = \dfrac{1}{N} \sum (x_i - \bar{x})^2$

Where: V – variance, N – sample size (number of individuals measured), $x_i$ – value of an individual, $\bar{X}$ – arithmetic mean.

## *MICROEVOLUTION*

Mutations occur at steady rates in human populations (Table **2**). In every generation some individuals will carry mutated genes and may express them in their bodies as anomalies. These anomalies are likely to affect their ability to survive or to produce offspring. In this way, mutations will not be passed on to next generations. This process of poorer reproductive performance of carriers of some genes is called "natural selection". And this is all there is to Natural Selection much maligned for its role in evolution. It is a simple process that eliminates defective genes from human populations and by the same token promotes spread of genes giving people advantage in reproduction.

**Table 2. Estimated rates of some mutations causing diseases in humans.**

| | |
|---|---|
| Albinism | 1 per 35,000 |
| Muscular dystrophy | 1 per 10,000 to 20,000 |
| Achondroplasia (a form of dwarfism) | 1 per 15,000 to100,000 |
| Microcephaly (abnormally small skull and brain) | 1 per 20,000 |
| Retinoblastoma (an eye tumour in children) | 1 per 50,000 |
| Tay-Sachs disease (infantile idiocy) | 1 per 100,000 |

Data from various sources, rounded up to simple numbers.

The process of interplay between mutations throwing in new genes and natural selection removing defective ones and promoting the few improving our fitness is continuous. It occurs every generation, actually every time a new baby is conceived, born, raised and eventually coupled with opposite sex to produce progeny. No wonder, then that human evolution is still occurring, being a natural, everyday process.

Only a mere 150 years ago less than 50 per cent of children born lived long enough to have families, the rest died prematurely [12].

Since then progress in sanitation, medical practice and general living conditions allowed 99 % of newborn babies to survive, grow up and establish their own families. This is undoubtedly a great benefit – practically every baby born can have a long and fruitful life. Every human life is precious and should be protected and extended as much as possible. There is, however, a price human populations must pay for realising their ideal. In the present-day good living conditions supplemented by excellent health care, operation of natural selection became strongly curtailed. This phenomenon is known as the "relaxed natural selection". Its obvious consequence is accumulation of mutations that in the past were decreasing human fitness. In the present-day environment enriched by sophisticated health services, these mutations no longer cause significant harm.

Many anatomical variations increased their prevalence in recent times. For example the condition, known as *Spina bifida occulta*, a birth defect of the lowest vertebrae overlaying the spinal canal carrying major nerves to the legs — has increased dramatically, growing from 11% of the population to 25% in 2000 years [78]. This trend of increase continued during the 20th century where prevalence of *Spina bifida occulta* increased significantly among patients of Australian [79] and Swiss hospitals [80].

Furthermore, during the 20th century the median artery of the forearm increased [81] its presence while that of the *Thyroidea ima*, a small branch of the aorta reaching up to the thyroid gland, practically disappeared [82].

The shape of the human braincase is numerically described by showing what percentage of its maximum length (from the forehead to the very back) is constituted by its maximum width. High width/length ratios are characteristic for round heads while low ratios describe elongated, ellipsoid heads. In the last 1000 years shape of the head, as expressed by this ratio – called a cephalic index – changed in Europe significantly, from about 75% to 83% [76] while it remained unchanged at about 73% in Africa [83]. These phenomena not only illustrate a recent change in the shape of a human body part, but also show that distribution of human variation across continents changes. About 1000 years ago head shapes of Europeans and of Africans were the same, now they are different!

At the end of the Ice Age (Pleistocene) about 12 thousand years ago, Europeans were as tall as they are today – approximately 1.75 m for males and 1.65 m for females [84]. With the advent of agriculture during the Holocene, some 10 thousand years ago, the food supply became poorer and monotonous, being based mostly on starchy plant products, people became settled in crowded villages, towns and cities that promoted spread of infectious diseases. All this led to a reduction in stature and body robusticity, so that in the times of the Roman Empire inhabitants of cities like Pompeii were about 10 cm shorter than at the beginning of the Holocene – men stood barely 1.65 m, females 1.54 m [85].

This kind of average stature persisted until the Industrial Revolution of the 18th and 19th centuries when living conditions for the urban proletariat (=working class) were even worse than in rural villages – poor nutrition, crowding, contagion, poor sanitation and lack of exposure to sunlight. It was only in the 20th century when living conditions started to improve and health became better that statures started increasing. By the end of the 20th century they were restored back to the same level they had before the beginning of the Holocene. This process runs at different rates in various countries depending on how situation of their citizens improved.

Australian females measured in 1926 by Berlei Ltd., stood an average 1.61m. Our own survey in 2002 showed body height of 1.62m, an increase of about 15mm, a smaller increase in stature than in many other countries. But the change in body weight was about 10kg [86]. In Europe, and in parts of United States of America, increases in stature were considerable, reaching to 100-150 mm per century [87]. Reasons for different rates of body height increases in various countries are still unknown since general improvements of living conditions in many of them were about the same.

Human bodies today are not adapted to natural conditions, human bodies are adapted to our cultural and technological way of life. We have changed the way people function and we have a happier and more productive society – due to a very substantial effort from medical science and public health measures.

## CHAPTER 8

# Living with Allurement: How the Universe Realises Itself

**Abstract:** The idea of allurement alludes to themes of creation and the centre of existence. In the previous chapters, human evolution has been explained. In this chapter, a grander outlook of evolutionary processes is explored. From cosmic beginings to human rituals, all are underpinned by allurement in all its kaleidoscopic magic and mystery. The ancients understood the power of allurement and coded it in all aspects of ritual behaviour and mathematical language. Allurement is the integument of existence.

**Keywords:** Amalgamations, Big Bang, Communitas, *Hajj*, Integration, Multiverse, Panspermia, Symbolic consciousness, Tardigrades, The centre.

## INTRODUCTION

In the previous chapters, several arguments have been presented to understand evolutionary processes in the human species. A fundamental driver of human evolution has been its unique kind of consciousness based on symbols. While symbolic consciousness is the hallmark of *Homo sapiens*, it is a veritable double edged sword. While conferring abstract thought, planning, music, empathy and art it invariably genuflects towards an array of pathological behaviours. Various theorists have proposed reasons for this psychological inconsistency in humans. One of the most compelling has been offered by the eminent biologist Paul Erhlich who points out that human nature is a product of biological and cultural evolution. This means that humans are not slaves to their genes, rather their nature coevolves between the biological and the cultural [88]. This is an on-going project of far reaching implications to both our species and to the earth.

Maciej Henneberg & Arthur Saniotis

Currently, the earth is undergoing a ninth major extinction event. The last major extinction event happened approximately 34 million years ago. While the earth has experienced eight major extinction events in the last 650 million years, the present one has been engineered by a single species – Us. The rapid extinction rate nature is currently undergoing is as much as 100 times the average rate of species extinctions. At this historical juncture, our species needs to understand more integral ways of living with the world. In this chapter, we argue that allurement not only plays an important part in the cosmos and nature, but is a way for understanding human concern with origins. Human need for understanding origination may play a vital part in our future evolution as we come to the realisation that we cannot escape the cosmic urge towards realisation.

## FALLING BY THE WAYSIDE

The issue of ecological degradation has been given ample scholarly attention. Prompted by Rachel Carson's controversial book *Silent Spring* (1962), the environment was centre-pieced as necessitating concern due to the profligate character of modernity [89]. Drawing upon Carson's thesis, future ecologists developed various theories of nature. Many works endorsed a transition from an anthropocentric worldview to an ecocentric worldview.

Developing from indigenous cosmologies, the famous ecologist Arne Naess eschewed from anthropocentric paradigms in developing his eco-philosophical model. Naess's model initiated a link for co-operating with nature with social justice, peace and non-violence. Naess was one of the first theorists to suggest that ecological degradation was tied to the failure of social systems in ensuring citizens' civil rights and social equality. In Naess's view, the 'othering' of nature is symptomatic of societies that have lost their way.

Humanity lauds itself over its mastery of nature. Human fixation for treating nature as 'Other' is prompted by a need for ontological security. Unable to admit our mortality and fragile existence, humans efface nature in order to retrieve a sense of existential mastery. However, our dominance is driven by distorted myths and commercial realities. Our cruelty must be absolute. Animals are tormented in order to fulfill market quotas, lands are denuded of their forests and animals are genetically modified for human consumption. In the words of Mary Jenkins, current animal husbandry is becoming more odious and morally reprehensible [90]. Jenkins continues stating that animal experimentation is akin to a war against the non-human world [90]. The implicit assumption here is in maintaining a mystique of human suzerainty in order to counteract our loss of control through age and death. Human addiction to "a deep cultural pathology" [91] diminishes any possibility for a rapproachement with nature. As Thomas Berry has stated, our assault on the earth has emasculated our experiential being. In our denial of nature we abjure the ties which are the source of our archetypal myths. The question remains, how far will human hubris inform our relationship with the earth? Our need for a universally moral response must outweigh our current medieval mindset [90]. Jenkins further notes that human separation from nature quintessentially embodies human hubris [90].

While extant humans in general show indifference towards nature, it is interesting that many core religious rituals and stories are somehow tied to nature. The humble non-human retinue of the Christmas nativity scene, the Buddha receiving enlightenment under a *bodhi* tree, or the 'night journey' of the prophet Muhammad upon the *Buraq* attest to the innate correspondence between the human and non-human worlds. For the eminent mythologist Mircea Eliade our most sacred rituals move us because they restore our place in the cosmos; they allow us to participate in the universe [92].

## THE 'OTHERING' OF NATURE: HISTORIC PROCESSES

Constructions of nature as other came during the Socratic period (circa 5th century B.C.E.). Unlike the pre-Socratic Greek philosophers (circa 6th century B.C.E.) (*i.e.* Thales, Anixamander, Anaxmenes, Heraclitus) who were among the first thinkers in natural sciences, the Socratic philosophers (Socrates, Plato, Aristotle, sophist schools) veered towards humanism. This transition is mainly attributed to the seminal Greek philosopher Socrates (470-399 B.C.E.). According to legend, it was Socrates who said that there is nothing to be learned from trees, and encouraged investigation into human nature. Socrates legacy was continued by Aristotle (384-322 B.C.E.). Although Aristotle made substantial inquiries into the natural world, he believed that human beings were superior to non-humans. This was based on his belief that human beings were rational animals. Spencer claims that Aristotle refutes animals with the ability to think, governed by instinct and needing human control for survival [93].

Such a view concords with Aristotle's idea of the inequality of beings where subservience is appropriate because it accords with the order of things [93]. Thereafter, with the exception of the Atomist philosophers, Greek philosophy was anthropocentric with little exploration into the non-human universe.

In contrast, the European medieval world consigned the universe to a sacred order from which humanity derived. The concept of a sacred universe, created and maintained by a Prime Mover (God), informed medieval cosmology. In this worldview (as prescribed by the Abrahamic scriptures), humanity was designated the role of vice regent; its purpose was to protect and preserve nature. Western religious iconography of the medieval period attested to the relationship between humankind and the cosmos. 'Man' was viewed as *'microcosmos'*, a theomorphic being composed of matter (earth) and spirit (*pneuma*). Humans embodied an "organistic unity" [94].

The preservation of nature was crucial, particularly in Islam as the former was a means for contemplating God's natural order. For the 'people of the book' (Abrahamic religions) nature was revelatory, a living testament of Divine design, composed of myriads of signs. Nasr explains that the aim of medieval science was to contemplate the unity of the cosmos as a way of understanding Divine Unity [95].

The advent of the European renaissance (1450-1550) fostered a paradigmatic shift. Renaissance science separated nature from the sacred, giving rise to a human centered universe [96]. Human allegiance to God was increasingly supplanted by humanist self interest [96]. The ecologically orientated space-time model of the medieval world was displaced by a linear scheme which adumbrated human living patterns according to economic imperatives.

The change in worldview was confirmed by various European thinkers. Among those, the French philosopher Rene Descartes's materialistic philosophy had a profound impact on western science. Descartes subscribed to the Aristotelian concept of an atomistic universe – a mechanistic cosmos stripped of its sacred substance. Moreover, his mind/body model reaffirmed the intellectual and moral supremacy of humans while constructing non-humans as soul-less machines. Here, Aristotle's concept of the 'Great chain of being' was employed in which the natural order created plants for animal subsistence and animals were created for human use (Politics 1:88) [97]. The assigning of animals to automata, bereft of feeling and emotion, resulted in their brutalisation by humans from the 17th century onwards. By this time, scientists indulged in animal vivisection. Scientists believed that animals felt no pain.

Accordingly, the historian Lynn White writes how progress in European agricultural technology changed humans' relationship with nature, where humans become master [98].

Another characteristic of the scientific worldview was a concern with taxonomy in order to manipulate the natural world for making it predictable [96]. For example, in his famous work *The Birth of the Clinic* (1975) Michel Foucault avers that physicians' increasing use of vivisection from the 16th century onwards not only revolutionised medical techniques but reinforced the notion of the 'body as machine'. This new medical 'gaze' advanced medical specialisation and its cavalcade of specialists for presenting scientific knowledge as an immutable truth [99]. Bodily disease was no longer viewed as a mystery but arising from pathological causes which could be carefully observed, defined and categorised by the new empirical discourse.

Along with the advent of a mechanistic worldview arose the discovery of the 'New World' by western Europeans. Colonialism spread western values to the conquered peoples of the 'Orient.' Europeans justified their conquest of the Orient on the pretext that they were a civilising force. Consequently, the colonised were viewed as being biologically, intellectually and morally inferior. Edward Said notes that the colonial powers essentialised the non-West as irrational, carnal, and inimical [100]. In this schemata, Oriental peoples were considered to be closer to nature. Thus, the 'Othering' of nature was similarly employed to define the Oriental. Nowhere is the imagery of Othering better exemplified than in Orientalist art of the 19th century. Orientalist art invariably portrayed the Oriental male as effeminate yet inimical, while the Oriental female was typically exotic and awaiting to be subdued [101]. The concurrent popularity of social Darwinism (circa mid 19th to early 20th centuries) described indigenous peoples as being imprisoned in irrationalism and were incapable of acquiring civilisational status. Like nature, indigenous peoples were to be controlled under the prescriptive of justified force.

The estrangement between the human and non-human other has been referred to by Martin Heidegger as a "night world" and Max Weber as "disenchantment of the world". For Heidegger modernity is consigned to instrumentalism whereby things are viewed technologically. Hence, things are no longer revealed as themselves but rather as resources for human ends. According to this scheme, old growth forests become chipboard, serene valleys become tourist resorts, and wildlife are turned to media attractions [102]. Modernity is characterised by a matrix of technological relations which has disconnected human beings from the earth and hurled them into existential limbo [102].

Similarly, Weber postulates that the social organising principles of modern human beings are essentially meaningless [103]. Disenchantment also describes the belief that the universe has become de-mystified and no longer perceived with awe, but succumbing to human calculations [103].

## THE "NEW STORY"

Catholic theologian Thomas Berry and cosmologist Brian Swimme have over several decades developed a new way of understanding they have termed the "New Story". The New Story expresses the need for a "mythic consciousness" [104] in humans which will ordain a rapprochement between the human and non-human worlds. Berry observes that Western science and religion have colluded in a global pathology which has vehicled a drastic gulf between the Western religious traditions and nature. According to Berry *et al.* [105], Western religious traditions have also been primarily focussed on human redemption while being critically deficient in recognising nature as a revelatory storehouse.

At this point, Berry argues that science must appreciate both poetic and spiritual elements of the earth [104]. Berry's views are synonymous with the eco-philosopher David Abram when he calls for a phenomenology of perception of the human 'presences' the non-human world. Here, the senses become increasingly attuned to the animal and organic landscapes, to the sentient cosmos [106]. Both Berry and Abram argue for a new kind of poietic embodiment emulating the mytho-experiential understandings of creation that are found in shamanic societies.

However, Berry is not calling for modern societies to return to a pre-industrial age, but rather a reinvention of the way in which humans idealise and participate with the earth. Concurrent with the New Story is viewing the earth as the ground of our biological and spiritual being. In Berry's *Dream of the Earth* he observes the human need to make the earth a primary source of human spirituality [107]. As humans are terrestrial creatures, our spirituality is earth bound. Such a worldview entails a transformation in regarding the earth from an insentient object to a spiritual parent [104, 108].

In this view, science becomes a key mode of interpreting the evolution of the universe and life on earth. The story of cosmogenesis begins with the primordial explosion of the Big Bang which is the origin of space/time, the creation of atoms and the first stars and galaxies [109] the beginning of our solar system and primeval earth, the advent of proterozoic creatures, the evolution of complex life forms, and the formation of human consciousness. As Swimme points out, defining life in a moral and spiritual sense only undermines biological life [110]. Importantly, Berry's new vision argues for a process based universe in which cosmic processes are disclosed by kaleidescopic creativity [110]. Berry's vision is laced with metaphors of an integrated universe where star matter is the progenitor of future worlds, and where life and consciousness are immanent principles of universal evolution [110].

Humanity is central to Berry's vision, as being the consciousness of the universe, a species created for celebration, for in human awareness the universe exalts itself [110]. Berry refers to the Chinese pictograph "*hsin*", in explaining the human relationship with the universe [111]. *Hsin* denotes the human as the "heart of heaven and earth". For Berry humanity has an opportunity to realise its integration with the earth and its ultimate realisation.

## THE ALLURING WAYS OF THE COSMOS

According to Berry and Swimme the cosmos is constituted by interdependent systems, or what he refers to as "allurement" where the cosmos is held together atomically, magnetically and chemically [112]. Allurement is the primal dynamism, the blueprint of cosmogenesis. This key principle which innervates macrocosmic evolution also informs human sociality [112]. For Swimme, allurement is cognate with desire. He notes that desire in the form of electromagnetic and gravitational interactions has informed cosmic processes since the beginning of time.

The recent discovery of a 13.7 billion year old universe is replete with the cosmic signature of allurement. Cosmologists tell us that the universe began as the size of an atom, doubling at an incalculable rate of 10-35 per second [113, 114]. This was the primordial Big Bang. Within a few microseconds elementary particles called leptons and quarks, the precursors of atoms were formed which in turn formed protons and neutrons [115]. The explosion that delivered the universe had varied by one part in $10^{60}$ – yes, one followed by 60 zeros – what we see around us would not exist and we would not be here to see it. Davies notes that if you wanted to fire a projectile at a one-inch target some twenty billion light years away, its accuracy would have to be at a quantum level [116].

At this stage the universe was devoid of structure and was composed of an opaque plasma. The elementary particles began to attract each other to form nuclei which in turn formed into atomic communities [115]. From the beginning it seemed that this universe was blueprinted for matter. Modern science has pointed out that the speed of expansion of the universe was critical to its existence. After half a million years as the universe began to cool the first elements formed – hydrogen, helium and lithium.

There has been recent speculation about the speed of the initial inflation of the universe. Some cosmologists have noted that even from the beginning cosmic forces had to be extremely finely tuned. The theory goes that for the universe to develop life there needed to be an extremely precise density on the scale of one part in 1015, otherwise the universe would have collapsed [117]. Similarly, if the expansion of the primordial universe was too fast by one millionth of one percent no galaxies or stars would have formed [115]. We are finding that the degree of density precision which was necessary in making the universe was as astonishing as the Big Bang itself. Similarly, the early universe needed to have a delicate balance between hydrogen and helium. If the force of gravity was too weak the universe would have consisted of 100% hydrogen which would have caused the newly formed neutrons to decay. If the gravitational force was weaker still all the neutrons and protons would have been bound up in helium nuclei which would have led to a 100% helium universe with no hydrogen to fuel star formation [118]. Unsurprisingly, intelligent design proponents have subscribed to the fine-tuned universe hypothesis. Challenging the intelligent designers, Linde has proposed a multiverse [119, 120] that is comprised of limitless mini-inflated universes that have a different "set of constants and physical laws" [120].

Parting from the fine-tuned universe debate, the allurement for matter to accrete in the early universe has been spurned by a recent discovery which pushes back the formation of galaxies to 200 million years after the Big Bang. A distant cluster of galaxies called Abell 383 would have among the first galaxies to have cleared up the hydrogen fog of the early universe in order to make ultraviolet light transparent when the universe was one billion years old [121]. Even after the first hundred million years referred to as the 'cosmic dark ages' the first star formations were humble in comparison to the giant star systems which came later on. In a laboratory experiment a team of scientists simulated the creation of a protostar. In their summary, the team noted that protostars of the early universe were the synthesis of gravity acting on gases and dark matter. In addition, protostars had the potential of becoming massive stars which could synthesise heavy elements [122].

The principle of universal accretion which had informed the early universe has revealed itself in various cosmic processes. According to Fathi, larger galaxies like the Milky Way galaxy may be the assimilation of smaller galaxies which have been dragged in by their gravitational force. These kinds of galactic fusions may be common to spiral galaxies which are more dynamic than elliptical galaxies [123]. The latter lack the ability for stellar generation and represent 'dead-end' galaxies. Such cosmic amalgamations could be interpreted in two ways; first, as a form of cosmic natural selection in which larger galaxies out compete smaller galaxies for resources (*i.e.* gaseous space existing between galaxies which is essential for star formation), or a kind of galactic symbiogenesis in which smaller galaxies have the opportunity to increase their creative potentiality. In this case, astronomers have hypothesised that galactic mergers are instrumental in inciting star birth, and represent an important part of cosmic evolution [124]. It seems that such mergers have been three times more common between large and dwarf galaxies than between two large galaxies over the last nine billion years. However, it seems that galaxies referred to as Brightest Star Galaxies (BSGs) which may contain up to 100 trillion stars and which were created by mergers ceased this inter-galactic fusion 7 billion years ago [125]. Scientists are baffled why this kind of activity had ceased approximately half of the universe's present age. Perhaps that this phase of galactic mergers was necessary in the early universe in order to consolidate future cosmic evolution.

## PANSPERMIA: IF WE ARE ALIENS

From Berry's and Swimme's model we derive a universe that is incessant in creativity, manifested in cosmic patterns of kaleidoscopic dimensions. Berry and Swimme remind us that our Milky Way galaxy is not only one of one hundred billion other galaxies, but that each galaxy's destiny is tied to all other galaxies [126]. As they state, Cosmogenesis is a form of organized communion. Allurement is tied into the fabric of the cosmos as some grand web of limitless proportions and beauty. Their approach entreats us to understand the inter-relationality of existence. Indeed, as they point out, even at an atomic level disconnection is an impossibility. All existence is connected. We are reminded here of the Mahayana Buddhist conception of the cosmos as encapsulated by the story of Indra's web. In this story the Vedic god Indra fashions a net in a heavenly abode stretching to infinity in all directions. At each vertex is a multi-faceted jewel; those jewels are also infinite in number. If one gazes into a jewel he/she can see all other jewels. This story is an elegant metaphor of cosmic inter-relationality which in Buddhism is referred to as *pratityasamutpãda*, which in Sanskrit means 'the together rising of all things'.

Let us look into this notion of inter-relationality further. If the universe is interrelational and the destinies of galaxies are tied, then it can be argued that galaxies may have the capacity of sharing information with each other. For instance, we know that the remnants of supernovae inform cosmic evolution. Another way in which the universe may be informed by informational sharing is panspermia. In a nutshell, the theory of panspermia explains that microbial life can be transported *via* bodies such as comets and asteroids to worlds of planets similar to earth where under ideal conditions they can grow and proliferate. The theory of panspermia (in Greek meaning "seeds everywhere") was first proposed by the Greek philosopher Anaxagoras (500-428 BCE). Anaxagoras also proposed the rudiments of atomic theory which became later synthesised by Democritus (460-370 BCE). His hypothesis was that seeds in the universe existed everywhere and that they could be spread *via* stellar bodies. Anaxagoras's hypothesis did not demarcate from more modern ideas of panspermia developed by Hoyle and Wickramasighe [127]. These authors had even hypothesised a controversial theory that some extant viruses and bacteria may be having a deleterious impact on human health.

While the panspermia theory has received some critical attention, recent findings seem to vindicate it. NASA scientists had recently duplicated the conditions of interstellar space by creating primitive cells that mimic life. The building blocks of these proto-cells can be found in interstellar clouds. These scientists believe that molecules essential to cellular membranes are immanent in the cosmos. The recent discovery reveals that genetic material (molecules present in RNA and DNA) found in space may have entered Earth's atmosphere embedded in meteorites, and therefore, was critical to the development of life on earth [128]. NASA scientists agree that early life may have had adapted nucleobases from meteorites in order to use in its genetic structure. If this is the case, then, life's raw materials could be widespread throughout the cosmos.

In 2009, NASA scientists reported that they had discovered the amino acid glycine in samples of the comet Wild 2 by NASA's sardust spacecraft [129]. This was the first time that an amino acid had been found in space. Director of the NASA Astrobiology Institute stated that: "The discovery of glycine in a comet supports the idea that the fundamental building blocks of life are prevalent in space, and strengthens the argument that life in the universe may be common rather than rare." To strengthen their case NASA scientists found that the glycine molecule had been composed of Carbon 13 isotope, which is of extraterrestrial origins [129]. According to Wickramasinghe *et al.* [130] the number of micro-organisms living in a comet can be exceedingly small. Interestingly, the warm interior of comets could provide a comfortable home for micro-organisms for a million years [131]. Comets may have the capacity of acting as hosts to extremophiles which can adapt to a comet's anaerobic and chemical environment. Such micro-organisms may also be oligotrophs thereby surviving in a comet's low nutrient environment.

The panspermia theory was given further support by an experiment exposing tardigardes in space in low Earth orbit in September 2007 by the European Space Agency (ESA). Tardigrades are microscopic animals with eight legs. There are some 1150 known varieties of tardigrades which can live in extreme environmental conditions. For example, tardigrades have been known to survive at temperatures -213°C [132]. Tardigrades are also eutelic, having a fixed number of cells when they reach adulthood. European experimenters took samples of desiccated species *Richtersius coronifer* and *Milnesium tardigradum* and exposed them to the space vacuum of low earth orbit. Astonishingly, some *M. tardigradum* survived solar radiation exposure of levels higher than 7000 kJm-2 [133]. Such tolerance to the space vacuum may be attributed to highly efficient repair mechanisms of damaged DNA [134, 135].

The tenacity for life to endure in extreme conditions and inordinate time frames was demonstrated in research conducted by biologists Monica Borucki and Raul Cano in 1995 where they had extracted bacterial spores from a bee's abdominal tissue. The bee had been entrapped in 25-40 million year old amber [136]. When the spores were placed in a suitable culture they were revived. The ancient bacterium was related to the extant species *Bacillus sphaericus*. The biologists were rigorous in their experimental procedures in order to prevent environmental contaminants [136].

Life on earth seemingly began in the form of microbial life approximately 4 billion years ago. These early prokaryotes would have adapted to the extremes of the Archean (3.9-2.5 billion years ago) environment. An analysis of sedimentary organic matter dating back to 3.8 billion years recorded an increased ratio of 12C to 13C – a sign of prolific microbial life some 500 million years after the formation of the earth [137]. However, there is a major problem with current theories of ancient life. Getting the basic building blocks of life to become nascent life is a stupendous hurdle of unimaginable proportions. The noted biologist Lynn Margulis poignantly notes that, "To go from a bacterium to people is less of a step than to go from a mixture of amino acids to a bacterium" [138]. Klyce puts it like this:

*"But if precellular life did that, it would need lots of time to create any useful genes or proteins. How long would it need? After making some helpful assumptions we can get the ratio of actual, useful proteins to all possible random proteins up to something like one in 10^500 (ten to the 500th power). So it would take, barring incredible luck, something like 10^500 trials to probably find one. Imagine that every cubic quarter-inch of ocean in the world contains ten billion precellular ribosomes. Imagine that each ribosome produces proteins at ten trials per minute (about the speed that a working ribosome in a bacterial cell manufactures proteins). Even then, it would take about 10^450 years to probably make one useful protein. But Earth was formed only about 4.6 x 10^9 years ago. The amount of time available for this hypothetical protein creation process was maybe a few hundred million or ~10^8 years. And now, to make a cell, we need not just one protein, but a minimum of several hundred"* [139].

Even if pre-cellular life had an abundance of elements and building blocks at hand to create life during the Archean era the appearance of life was too short. Moore articulates his concern about this time problem as follows:

*"Of one thing we can be certain: The RNA world—if it ever existed—was short-lived. The earth came into existence about 4.5 x 10^9 years ago, and fossil evidence suggests that cellular organisms resembling modern bacteria existed by 3.6 x 10^9 years before the present. There are even hints that those early organisms engaged in photosynthesis, which is likely to have been a protein-dependent process then, as now. Thus it appears likely that organisms with sophisticated, protein based metabolisms existed only 0.9 x 10^9 years after the planet's birth"* [140].

In other words, there must have been an interval period in the biosphere of less than 100 million years from the Earth's beginning in order for RNA based life forms to emerge, an improbable likelihood since during this time the earth would have been bombarded by meteorites, making the planet uninhabitable [140]. When thinking along these time frames, the emergence of RNA life forms, let alone, the transition from an RNA to a DNA dominated world which requires many stages seems dubious [141]. For this reason, the possibility of panspermia may provide a tenable alternative considering the minute opportunity for life to form [140].

## RETURNING TO THE CENTRE

Continuing our journey of allurement we can observe its manifestations in the notion of the centre, an archetypal idea which has informed the human psyche since prehistory. Jung referred to the centre in terms of the integration of the psyche, the point of psychological origin. He referred to the psychological centre as the collective unconscious which was universal to the human species. The collective unconscious was an artifact of the human mind stemming back from our non-human ancestry. It is difficult to ascertain to what extent our human ancestors had been informed by the collective unconscious. Prehistoric shamans were able to enter into altered states of consciousness using various ritual techniques. These shamans were the first people to have willfully crossed the threshold into domains of unexplored mind. Their regular journeys into the deepest layers of the psyche probably informed the evolution of consciousness of modern humans by integrating disparate areas of the brain [142]. To understand how prehistoric shamans would have manipulated the psyche we need to look at extant shamanic cultures. Present day shamanic cultures use music, dancing, sensory deprivation, sensory stimulation, fasting, and psychotropic substances to induce altered states of consciousness. The various ecstatic techniques used in these cultures may be viewed as stimulating neuro-peptide communication within the brain and body towards achieving healing within the body. The techniques induce a neurochemical cocktail within the brain and body that may enhance healing. For example, symbolic manipulations are significant in shamanic medicine. In addition, manipulation of emotions during ecstatic techniques has physiological consequences. Opiates produced in the limbic system (focal point of neuropeptide receptors) which regulates emotions may be seen as bio-feedback loop where opioids inform affective states and affective states inform opioid release [143]. In other words, humans seem to be hard wired towards journeying psychologically into themselves in a way that is precluded in other species. Our very being is informed by a need to return to the centre.

Many cultures have a notion of the centre which informs their rituals and myths. The origins of the centre may go back to early humans over one million years ago. Early humans would have noticed that the sun acted as a kind of centre which regulated the diurnal patterns and seasons. It is difficult to ascertain how stone age humans idealized the sun. Perhaps, they noticed that there were two centres, the sun governing the day and the moon governing the night. They would have also noticed how during intermittent periods these two centres coalesced during lunar eclipses. Such events must have evoked both awe and ambiguity. Unfortunately, we know little about how upper Paleolithic humans (~40,000-10,000 years ago) envisaged the sun and moon in their mythopoeic imagination. However, the famous figurine called Venus of Laussel dated from the Upper Paleolithic period (25,000 years ago) depicts a stylized nude female figure holding a wisent horn. The horn has 13 incised notches. According to some researchers the notches may symbolize the number of lunar or menstrual cycles in one year.

According to the anthropologist Michael Winkelman prehistoric cave art depicts profound mythological elements which scaffold the mind and express an innate human desire with the centre [144]. This is also concurred by Clotte & Lewis-Williams who profess for a neurologically based art based on shamanic practices. These authors suggest that early human art was profoundly predicated on mystical experiences triggered by shamanic practices such as sensory deprivation or use of hallucinogenic substances. In earlier and extant hunter-gatherer societies shamanic practices are universal and manipulate the cartography of the human mind [145]. What is evident in shamanic rituals is the idea of entering into the inner centre, as a method of integrating modes of consciousness [144]. In other words, the outward depictions of centres as depicted in art and technology reflect integrative elements of the human brain. Interestingly, the human brain seems also to have a propensity towards hypnotic states which may have assisted early humans in diminishing stress responses, pain reduction, stimulation of the immune system [146] and increasing social bonds during ritual activity [147].

Returning to early cave art, it is apparent in some figures that they may have been painted during mystical experiences. A fascinating aspect of Cro-Magnon cave art sites is their veritable inaccessibility, which may have heightened their luminal quality. The artwork of some caves can only be reached by going through long, narrow tunnels or passages. Some of these passages are small enough to just enable a person to squeeze through them. The sensory deprivation combined with the arduous effort in reaching destined cave art sites may have been an intentional device for disorienting individuals, whereby heightening feelings of awe and mystery.

All these complicated hidden passage ways lent themselves to extraordinary effects which would be inexplicable to uninitiated novices, who must have been deeply impressed. The effect of songs, cries, or other noises, or mysterious objects thrown from no one knows where, was easy to arrange in such a place [148].

Cro-Magnon cave art represents a human encounter with profound religious and mystical symbolism that may have induced altered states of consciousness [149]. Additionally, the various artistic representations apparently correlate with an inward journey characteristic of altered states of consciousness [149]. Animal representations apparently followed similar gendered features. This aspect of female sexual symbolism seems to be stamped within the Lascaux cave located in France, "which would explain the care with which narrow passages, oval shaped areas, clefts, and smaller cavities are marked in red, sometimes painted entirely in red" [148]. The feminine symbolism with the Lascaux cave is complemented by the ithyphallic presence of the costumed shamans who may have overseered the initiation ceremonies [149]. The manifestation of female and male symbolism may also represent a cosmogonic journey to the source of life. This notion of the inward journey towards the existential source of life is cognate with Eliade's notion of the major aim of ritual which is to return to the sacred and eternal order in order to achieve a sense of unity with the divine.

For the shaman unites in himself the two contrary principles; and since his own person constituted a holy marriage, he symbolically restores the unity of sky and

Earth (the world parents), and consequently assures communication between gods and men [150].

The idea of the centre became clearly manifest during the Neolithic period. During this time, humans began to congregate in settlements which prompted the construction of buildings and megaliths. The Neolithic period also saw the advent of the institutionalized religion. Like their ancestors, people during this time were very interested in religious matters and began constructing various kinds of religious structures which blended their understanding of the solar and lunar cycles and religious experience. Stonehenge and other monuments like it which are spread throughout the United Kingdom are typical of this movement. Stonehenge which was constructed between five and four thousand years ago has been at the centre of many disparate theories. The earliest traces of Stonehenge date back eight thousand years ago. In its heyday it must have been an impressive structure. The present structure consists of two concentric circles. The outer circle consists of seventeen remaining pillars while the inner circle has been reduced to six lintels. Many of the stones are colossal. One thing is for certain, the creators of Stonehenge had a well developed understanding of solstices, equinoxes and other celestial matters. Stonehenge probably had multiple functions which may have included curative, burial and shamanic based rituals. Stonehenge may also have been linked to other sacred sites.

An insightful theory proffered by John Michell considers Stonehenge to be a cosmic temple whose physical dimensions correspond with the vision of the holy city in the Book of Revelation [151]. For instance, Michell correlates various Stonehenge measurements with Christian sacred script based on Greek *grematria* where each letter of the Greek alphabet is consigned with a numerical value that assisted in establishing a connection between one word or phrase and another. This system of numerical formulae was pivotal to the sacred canon of many ancient societies, and was at the basis of many sacred buildings, monuments, and churches *i.e.* the Egyptian pyramids, Glastonbury Abbey, Chartres cathedral in France. This belief in alphabetical letters as correspondences of creational powers has probably been influenced by the Jewish *kabbalah* [152]. The *kabbalists* consider the *aleph-beth* as a distillation of the sacred [152]. In both systems, the cosmos reaches its zenith in the human form [153].

It is no coincidence that a derivative of *gematria* is geometry, the later being considered as a perfect science by the Greeks. Geometry was considered as being independent of the world, being aeonial, thereby, enabling the human mind to attain transcendent space [154]. This idea was central to neo-Platonists who continued the Pythagorean tradition, which later influenced early Christendom. The system of *gematria* is steeped in mystery. The early gnostics and church fathers used *gematria* in their mystical doctrines. Augustine believed that numbers embodied God's thoughts, and that the material and spiritual worlds were predicated on eternal numbers. The ancients were seemingly assiduous in trying to find all manner of correspondences between the material and spiritual worlds in order to reaffirm their connectedness. *Gematria* confirmed this inter-relatedness. For example, the numerical correspondence of the word 'Jesus' which in Greek is *Iησους* = 888. This number is closely related to the numerical correspondences for the:

'holy spirit', *το αγιων πνευμα* = 1080.
'fountain of wisdom', *πηγη σοφιας* = 1080.

The number 1080 is considered to be related to the moon's influence and is connected to intuitive modalities and the unconscious mind [151]. The oppositional number of 1080 is 666 which represents the sun, the intellect, and worldly matters. In contrast 1080 represents spiritual matters [151]. In closer scrutiny this number is connected to a plethora of phenomena. The philosopher Heraclitus noted that civilization is destroyed in fire every 10,800 years, while a quarter of the age of the Hindu Kali-Yuga (430,000 years) is 108,000 years. Other interesting connections include 108 beads in the Buddhist and Hindu rosary and 10,800 bricks contained in the Hindu fire altar. The average number of breaths in one hour is 1080 [151].

Ever since the famed philosopher Pythagoras, Greeks believed that the universe could be understood by numerical correspondences, a belief that is central to modern day numerology. Ancient Pythagoreans took numbers very seriously and swore their oaths by the *tretactys*, a triangle consisting of four rows which added to the number ten, a mystical number. The *tretactys* represented the four elements, earth, water, air, fire, and the harmony of the spheres. Moreover, the heavenly spheres moved according to mathematical principles [155] something which was verified by Newton's *Principia* (1685-1686). Both Pythagoras and Newton obsessed in finding mystical correspondences in nature. Newton spent many years pursuing his secret love of alchemy, a practice which he believed had been rooted in the Hermetic tradition from ancient times [156].

The Pythagorean notion of numbers spread to the Muslim world during the eighth century where it became employed in the form of magical squares by the medieval alchemist Jãbir ibn Hayyãn. He was renowned for his study in magical squares in which he incorporated complex mathematical and medical formulae. By using alchemical lore Jabir maintained that number is immanent in the cosmos. Jabrian theory categorised matter according to four elements: heat, cold, moisture and dryness. These elements were consigned to the numbers 1, 3, 5 and 8 [157]. A popular form of magical square depicts the nine numerals, from 1 to 9. The numerical values of each horizontal, vertical and angular sides equal the number 15.

| 4 | 9 | 2 |
|---|---|---|
| 3 | 5 | 7 |
| 8 | 1 | 6 |

This magical square is also traditionally attributed to the planet Venus who was called Ishtar in the Middle-east. Mesopotamian gods were often designated with numbers, and the number of the goddess Ishtar was 15 [158].

Michell claims that the builders of ancient churches based their architecture on sacred geometry, whose knowledge was later lost. Keys to this knowledge are to be found in the Book of Revelation concerning the dimensions of the New Jerusalem that will be established during the end times. A quick examination of the measurements of the New Jerusalem finds that it reflects cosmic dimensions. For example, the number 6 figures prominently which is the number of the cosmos in Greek, symbolizing cosmic order [151]. In his work *City of God*, Augustine wrote on the perfection of the number 6 for "in this did God make perfect all his works" [151].

In nature, the number 6 is found in the following:

Sun's diameter = 864,000 miles (12 x 12 x 6000).
Moon's diameter = 2160 miles (6 x 6 x 60).
Speed of the earth around sun = 66,600 miles per hour.

The numbers 144, 666, 864, 1008, 1080, 1224, 1728, 7920 which are multiples of 6 are also prominent in the dimensions of the New Jerusalem.

Returning to Stonehenge, when the New Jerusalem is superimposed on the Stonehenge plan they are almost identical. When the perimeter of the square of the New Jerusalem being 31,680,000 feet or 48,000 furlongs is reduced to 316.8 feet this number calculates to the perimeter of the Stonehenge sarsen circle [151]. Stonehenge represents a cavalcade of mysteries and is one of many ancient buildings found throughout the world which have yet to reveal their numerical secrets.

Of course number magic is predicated on measurement units used. When a metric system is applied the magic evaporates. Clearly a creation of the human mind.

## PERSONIFYING THE CENTRE: RITUALISING ALLUREMENT

As the dimensions of ancient buildings point to the correspondence between heaven and earth, human ritual life is replete with the notion of the centre. Jackson's idea of the metaphorical correspondences between different bodily domains is pertinent to this analysis since he argues that any action in one sphere will influence another domain [159]. He goes on to say that this possibility is realised whenever rituals are carried out in one accessible or disturbed domain that is considered analogically linked to it [159]. This idea is particularly important in relation to understanding rituals and how they attempt to inform ritual participants that what they are doing is out of the ordinary.

At this point, it is apt to introduce Victor Turner's concept of *communitas* in order to unpack some of Berry's and Swimme ideas on allurement. Central to Turner's theory of religion is the phase during religious rituals which he calls *liminality*. Turner had reformulated this idea from Arnold van Gennep's notion of *limen*. Gennep was one of the first anthropologists to make a focused study of human rituals. Gennep found that most rituals in non-western societies had a tri-partite structure; separation - where ritual participants were physically separated from non participants; limen - the second stage of a ritual characterized by participants' loss of former status; re-aggregation – where participants re-entered society conferred with new status and responsibilities [160]. Later on, Turner had found a similar phenomenon occurring in his study of ritual life amongst the Ndembu tribe in Zambia [161, 162]. However, Turner demarcated from van Gennep by associating the *limen* phase of ritual (which Turner re-worded as *liminality*) with some interesting ideas. Firstly, Turner suggested that *liminality* is characterised as an ambiguous state in which ritual participants are thrown in a symbolically statusless domain [163]. Turner considered liminality as being a sacred condition where individuals were symbolically maneuvered in opposite direction from everyday reality. This phase is also characterised by what Turner refers to as *communitas* which denotes a feeling of unity, comradeship, egalitarianism, empathy, and comity between individuals. In other words, *communitas* suggests a temporary dissolution of social structures which privilege social distinctions between people. For example, Turner asserts that religious pilgrims often experience a state of subjective unity unalloyed by social divisions and inequalities [164].

Turner cites that *communitas* is not restricted to rituals but also encompasses various kinds of ideological movements or transcendental states of awareness where an individual experiences a feeling of unity between human or non-human others. This corresponds with Martin Buber's "I-Thou" where humans are aware of the human or non-human other "as having a unity of being" [165].

Pilgrimage also seems to characterise *communitas* at various levels. According to Turner, pilgrimage embodies "normative *communitas*"; it is bound up by a set of conditions and religious codes which require a kind of fellowship or brotherhood among participants. As Turner points out:

The kind of communitas desired by tribesmen in their rites and by hippies in their "happenings" is not the pleasurable and effortless comradeship that can arise between friends, coworkers, or professional colleagues any day. What they seek is a transformative experience that goes to the root of each person's being and finds in that root something profoundly communal and shared [164].

For instance, the Muslim pilgrimage called *Hajj*, which is one of the five pillars of Islam attempts towards exhibiting *communitas* between Muslims. Firstly, Islam endorses its followers to regard the *Hajj* as a significant spiritual undertaking marked by physical and emotional trials. Upon entering Mecca a pilgrim must give a solemn intention to assume a state of purity. The pilgrim is forbidden to kill any living creature; they must not uproot any plant. No jewellery must be worn or perfume used, males are not permitted to shave or cut their hair. Male pilgrims must wear the *ihram* dress consisting of two pieces of white unsewn cloth. The ihram is supposed to symbolise the brotherhood and equality of all Muslims. Women are also permitted to wear white robes, but are not allowed to cover their faces [166]. The focus of the *Hajj* is the building called the *Ka'ba*, a cuboid shaped building covered in a black shroud (*kiswa*) embroidered in verses of the Qur'an. The *Ka'ba* is the focal site of prayer for 1.6 billion Muslims five times a day. The idea of the centre is ubiquitous in Muslim societies and is symbolised by the *Ka'ba*, being the most sacred shrine in Islam.

The *Ka'ba* at Mecca is a prime example of Eliade's notion of a sacred centre. For Muslims, the *Ka'ba* is considered as the sacral point of the world's beginning, and the nexus between heaven and earth [167]. Eliade exerted a profound influence on the study of pilgrimage by forming a notion of the pilgrimage shrine that becomes an archetype of the sacred centre. Such centre distinguishes itself sharply from surrounding profane space. This author describes how sacred centres are linked to the archetypal imagery. Worshippers believe that sacred centers are locations where creation happened. They are symbolic focal points of the universe wherefrom divine spreads over the world [92]. As Eliade points out, the centre is the point in which existence becomes impregnated by the sacred, emerging as a distinct entity yet dependent on the eternal and sacred other [92]. Victor Turner and other theorists have elaborated the concept of the centre introduced by Eliade. Turner thought that the point of intersction between profane and sacred is located in the pilgrimage shrine [168]. Additionally, the *Ka'ba* is viewed as the first thing created, some two thousand years before the creation of the Earth which was spread around it [167]. Muslims believe that the *Ka'ba* is erected on the spot where the first temple to the Abrahamic god was buit by the patriarch Adam. The foundations of this sacred house have been forged from the seventh earth by God's angels [167].

Barbara Meyerhoff's study of the Huichol Indians is significant as it re-formulates *communitas* as a universal communion between the human and non-human world. The Huichol perform sacred journey to *Wirrikuta* in the south-west United States of America. In Huichol cosmology *Wirrikuta* is the birthplace of creation and the Huichol deities. During one of their rituals the Huichol attain a transcendental state synonymous with Eliade's *illud tempus* where they unite with the gods and the cosmos [169]. To return to Wirrikuta is no less a return to the original state of paradisiacal existence, which in Eliadian theory is a panhuman desire for re-establishing the primordial condition of humanity [170]. As a way of reinforcing the sacrosanct nature of Wirrikuta the pilgrims incorporated reversal such as doing everything backward and speaking in a metaphoric language. These symbolic reversals provided a template for the unification of opposites, in order to create a condition of oneness and unity with all things. Secondly, in order to further reinforce the feeling of *communitas* the pilgrims perceived themselves as living incarnations of the Huichol deities. This was accomplished by symbolically discarding their human condition through ritual cleansing and becoming their 'divine counterpart' [169].

Symbolic reversals are an interesting ritual phenomenon practiced in many cultures, and are used in order to reinforce that a specific behaviour is out of the ordinary. Liminality, as noted earlier is a kind of symbolic reversal. It seems that doing things in a backward or counter-intuitive manner reaffirms to ritual participants a feeling of awe and attraction described by Rudolph Otto as *mysterium tremendum et fascinans* ('fearful and fascinating mystery') [171]. Otto thought that *mysterium tremendum et fascinans* enabled individuals to be in communion with the sacred other. Noted examples of this are:

1. The Christian Eucharist where through the process of transubstantiation the wine and wafer are turned into Christ's blood and body.
2. During the *Hajj*, pilgrims circumambulate the *Ka'ba* (*tawaf*) counter-clockwise.
3. Shamanic androgyny found in indigenous Asian, American and African

societies.

## ALLURING OUR BEING IN THE WORLD

As we have seen, allurement is a way of understanding complex cosmic, biological and social processes. The distant formation of matter as stars and galaxies was imbued with the same impulse as pilgrims seeking to be united with the sacred other. The lesson of allurement is significant. Place bacteria together and they will forget any qualms of individuality. They will act as a tightly knit phalanx [172]. The incessant movement to accrete drives cosmic and social evolution. The ancients postulated relationships within nature and the cosmos based on sacred numbers. Their buildings were thresholds between the living and the sacred realms. Sacred structures were outward centres pointing to the inward centre. As the Big Bang is believed to have derived from a single point within infinity, sacred centres remind participants that the centre is everywhere. Berry's and Swimme's New Story concerns itself towards a retrieval of the inter-relationality of life on earth. Their message is simple, view life on earth not as a collection of objects but as a communion of subjects.

With the exponential rise of technology the cosmos has engendered new organising principles. The advent of the internet and global communication networks have catapulted human consciousness. For the first time in history human consciousness has been transformed into a "Noosphere" enveloping the earth in an infinitesimal communication matrix. Here, a new kind of allurement has come into being, heralded by the symbiosis of silicone based cyber machines and biologically based consciousness that are extending ontological boundaries. The Christian scientist Teilhard de Chardin predicted that the increase in human complexity amplifies the noosphere. In Teilhard's teleology the noosphere will reach such a level of integration that it will culminate in the Omega Point – the apex of consciousness [173]. This idea aligns with Berry's claim that human consciousness is the integument which unites humans to the world and to each other [174]. It is conceivable that this is what the Sufi philosopher Muhiyuddin Ibn 'Arabi (1165-1240) referred to as *wahdat-ul-wujud* (Unity of Being) which viewed creation as constituting a unity. In the tradition of the Heraclitian idea of cosmic flux, Ibn Arabi states that the universe is continuously being created (*al-*

*khalq al-jadid*) [175]. Creation of the universe undergoes constant transformation like in a kaleidoscope. This includes the sphere of thought and the imaginary reality [174]. At each moment there are infinite possibilities of change, evolution and exploration [174].

# CHAPTER 9

# Other Ways of Knowing: Retrieving the Shaman in Us

**Abstract:** In this chapter, we discuss the realm of altered states of consciousness as a significant aspect of human evolution. We argue that there is a need for humanity to retrieve its 'shamanic' legacy as a way of answering current ecological challenges. Both shamanic socieities and their engagement with altered states of consciousness, and mind/body practices of Western societies are examined. Far from being irrelevant, we show that intuitive modes of knowledge provide unique ways for understanding the nature of human existence and the non-human world.

**Keywords:** Akhasic field, Feyerabend, Lycanthropy, Posthumanism, Shamanism, Sufi, Transhumanism.

## SCIENTISM'S MARCH

Whereas allurement is a central way of understanding our place in nature and the cosmos, in this chapter, we argue that humanity needs to retrieve polyphasic states of awareness in order to realise their bond with the non-human world. Polyphasic states of awareness refer to non-ordinary consciousness; dreams, visionary and mystical states, hallucinations, intuition, meditation, trance states, self-hypnosis, and near death experiences. These kinds of states are often referred to as altered states of consciousness. Polyphasic states of awareness offer other ways of knowing with the non-human world that are integral for our rapprochement with it, and foster new kinds of creativity. Furthermore, polyphasic states are needed in order to foster new kinds of knowledge to finding solutions to the ecological and social crises currently embroiling our planet. In a world bedevilled by systemic problems, polyphasic states offer a new way for experiencing and understanding the life world. Indeed, these states, as will be discussed are our evolutionary legacy.

**Maciej Henneberg & Arthur Saniotis**

It would be to our detriment if we negate or nullify their existence. Since prehistory, polyphasic states of awareness have been used as ways of eliciting knowledge in numerous hunter-gatherer and traditional cultures. These ways of knowing probably expedited human cognitive evolution as characterised by various art and ritual genres that are central to human cultures. How widespread are polyphasic states of awareness? Back in the 1960's the anthropologist Erika Bourguignon made a survey of the prevalence of polyphasic states. She discovered that out of 488 human societies 90% had practiced some form of altered states of consciousness [176].

The use of polyphasic states of awareness in non-western societies is in contrast to western societies which foreground monophasic or ordinary states of awareness based on rationality and empiricism. The interpretive power of scientism and technology has either relegated or dismissed polyphasic states to the realm of irrationality and fairy tales.

The historical pre-eminence of rational modes of awareness begins with the Greeks. It was Socrates who stated that nothing can be learnt from trees. From Socrates onwards there was an onus on humanism and logical ways of knowing the world. Socrates represents a historic shift from the oral tradition of the Greeks expressed by the Homeric traditions to more rational understandings of the world. Of course, this movement had already been developed by the naturalist philosophers Thales, Anaximander, Anaxagoras, Protagoras and Pythagoras in the fifth and sixth centuries BCE. However, Socrates was not particularly interested in the natural world but rather focussed on human intellection.

His student Plato further excluded non-rational modes of awareness by dismissing poetry and story-telling as being poisonous to the nobility of mind. Aristotle further privileged rationality by claiming that vision was the most verifiable of the senses since it gave a 'true' picture of reality. It is safe to say that Aristotle was sceptical about whether the other senses could report the world in a certifiable way. Aristotle encouraged empirical observation of the natural world. An indefatigable natural scientist, he proposed that nature is open to scientific inquiry. The Aristotelian tradition influenced the scientific renaissance from the 16th century onwards. For instance, Francis Bacon (1561-1626) foregrounded the inductive method. An avid empiricist, Bacon was known to have stuffed a chicken with snow and observed to what degree its flesh was preserved [177]. In his *Novum Organon*, Bacon described the logic behind the scientific method wherein empirical observation was central [177]. From the enlightenment period (17th century) onwards, scientism slowly supplanted religion as an explanatory model for existence. Newton's 'clockwork' model of the universe described the cosmos as adhering to mechanistic principles that could be observed and measured. Furthermore, Charles Darwin's seminal work *On the Origin of Species* (1859) unequivocally argued for an understanding of life on earth based on the scientific method. Darwin's work catalysed the life sciences and promoted a way of thinking which steered away from 'irrational' and unverifiable propositions of religion.

A lot has been said about the pre-eminence of science and its ability to reveal 'truth'. The noted scientist Richard Dawkins has been particularly vocal about non-rational ways of knowing, subscribing to a scientific fundamentalism. Dawkins scepticism reflects the inability of the scientific establishment to understand that their very science is also informed by non-rational modes of knowing – hunches, guesses, and 'magical fabulations' [178]. The iconoclast Paul Feyerabend went further. He proposed that there is no scientific method, and secondly, that science also relies on mystical ideas, intuitions, bluff and propaganda to concoct stories of its superiority over other ways of knowing [177]. Feyerabend was enthusiastically scathing in his attack on science in *Against Method* (1975), noting that scientific method was no more substantive than pseudo-science. Feyerabend had made his point – the scientific paradigm is not the *sine qua non* of human achievement but just another interpretive model. Keeping this in mind, important modern thinkers such as William James, Gregory Bateson, Ervin Laszlo, Stanislav Grof, David Abram and others maintain that non-scientific ways of knowing offer valuable perspectives for understanding the human condition and the non-human world. The prominent Indian activist Vandana Shiva contends that the scientific paradigm has systematically engineered itself to be the dominant discourse in understanding reality, while at the same time negating age old systems of knowledge. The latter have been comfortably consigned as inferior and representing a primitive method for accessing knolwedge [179]. On this theme it has been claimed that western based science is superior way of knowing the external world, whereas traditional cultures are comparatively inferior in this area [180].

Due to the scientific emphasis on rational knowing, and especially of physicalism—the belief that "mental entities, properties, relations and facts are all physical"—other ways of knowing, including intuitive knowing, have been regarded as an inferior epistemology at best and a vestige of superstitious nonsense at worst [181].

Of course such an attitude is itself problematic as the scientific paradigm has been often informed by non-empirical modes of thought. For example, the Nobel laureate Otto Loewi discovered the method of chemical transmission of nerve impulses in a dream. Similarly, Friedrich Kekulé (who will be discussed later) dreamt of the Benzene molecular structure.

Dawkins rationalises that non-empirical ways of knowing are fraught with speculation, self-deceit and are downright false. No doubt his analysis is correct in relation to a litany of charlatanry which he and others have observed. However, a similar claim can be held against science which has changed from a laboratory based endeavour to big business. One only has to look at the world of scientific patents or genetically modified foods. Big business has changed science. In the next sections, we will journey into the fascinating and surprising world of polyphasic ways of knowing.

## THE SHAMAN'S WAY

In order to understand polyphasic ways of knowing, we must explore their shamanic roots. Arguably, shamanism is the first known mystical complex in the human species. The term 'shaman' derives from Tungus culture of Siberia meaning 'one who knows'. The anthropologist Michael Winkelman identifies shamanism as the first neuro-theology [182]. Shamanism has its origins in prehistoric times. We know that shamanism had already existed during the Upper Paleolithic period (40,000-10,000 BCE). The cave paintings at Lascaux (France) and Altamira (Spain) suggest shamanic influences. Shamanism is tied to the wisdom traditions of ancestral and extant hunter-gatherer cultures. It is based on the idea that there exist various kinds of spiritual powers which influence the human world. Shamanism uses a repertoire of techniques in order to trigger altered states of consciousness during communion with the spirit world. Shamanic practices often employ ritual for manipulating the human psyche through symbols and metaphors which may prompt healing.

In the recent years, there has been a lot of interesting research conducted on the neurobiological elements of shamanism and why it may promote healing. One school of thought maintains that shamanic techniques may assist in integrating neural pathways [182]. It seems that the human brain is hardwired for multiple experiential phases [183]. The physiology of ecstatic states involves the auto- nomic nervous system [182]. Altered states of consciousness synchronise the limbic system which gives rise to emotions with the frontal cortex. During such states, the sympathetic system which regulates autonomic functions, such as heart rate,breathing and diaphoresis, is initiated and continues until the individual is over stimulated. This process then triggers the parasympathetic system (responsible for resting functions) inducing a relaxation effect that is characterised in trance and deep meditational states. Altered states of consciousness also stimulate the production of the 'relaxation' neurotransmitter serotonin [142]. Ancestral humans would have discovered that certain activities led to sympathetic and parasympathetic arousal. These would have included dancing, hyperventilation, drumming, chanting, fasting and sensory deprivation. An interesting feature of all of these techniques is that, they foster hypnosis in humans. Recent research points to the human propensity to easily enter into hypnotic states. Approximately, 90% of humans can bring on self-hypnosis. It seems that this hypnotic ability is found throughout the animal kingdom. Like us, non-human animals can induce hypnosis *via* ritualised and repetitive behaviours. Inducing self-hypnosis may have served various survival functions in ancestral humans over long periods of time so as to influence the "frequency of hypnotisability genotypes" [147]. Prehistoric rituals were probably used for releasing endogenous opiates that stimulated the immune system, reduced pain and stress, improved coping skills and fertility and strengthened group ties [182].

A central feature of shamanism is the soul flight or soul journey, where the shaman's soul separates from his/her body and journeys to other realms. The purpose of the soul journey is to commune with the spirit or animal powers, and is also used as a precognitive technique for determining lost objects or to explore the fate of human members. Oftentimes, the soul journey transports the shaman to the ancestral realm where he/she gleans esoteric knowledge to be used in the shaman's repertoire. The soul flight represents a transformation of consciousness whereby the psyche is integrated. During the soul flight, a shaman may encounter spirit helpers in the form of human and non-human animals. These spirit beings play a crucial part in guiding the shaman's soul towards attaining mystical knowledge. The soul flight is the psyche's adventure into the unconscious, and is integral to shamanic experience.

## THE DRAGON'S PSYCHE: EXPLORING NON-LOCAL MIND

The element of non-local mind which is a shamanic legacy has been reported in hundreds of past and present cultures. Non-local mind can be defined as the ability of one mind to influence another mind from a distance, or knowing something which is beyond the grasp of sensory perceptions. Many languages have vocabularies which describe non-local mind. In English, non-local mind is referred to as intuition, premonitions, déjà vu, telepathy, 'gut feeling', precognition, distance healing, remote sensing, extra-sensory perception (ESP), psychokinesis and clairvoyance. In Arabic, non-local mind is known as *kashf,* while in Sanskrit it is known as *tanmatras.* This word relates to the subtle sensorial perceptions which are implicated in higher states of consciousness. The telepathic communication has been believed in worldwide for millennia. Some indigenous cultures also believe in telepathic communication between human and non-human animals. Such communication is often mediated (but not exclusively) by dreams. Extant shamans are believed to be able to communicate with various non-human animals in both ordinary and altered states of consciousness. Throughout the Amazon region, shamans have the putative ability to communicate with various spirits, animals and plants [184]. For example, Native Americans often refer to plants such as peyote and san pedro (*Trichocereus pachanoi*) as mentor plants and believe that they have a telepathic affinity with these plants. These plants are believed to communicate with shamans of their potential healing qualities.

There have been approximately 900 studies conducted over the last sixty years on non-local ways of knowing [185]. Many of these studies were conducted in order to ascertain whether observers could affect matter [185]. The majority of these studies provide "independently replicable evidence that observers can affect the behaviour of physical systems" [185], thereby, supporting the probable existence of non-local mind. The sheer volume and veracity of these studies challenge cherished scientific myths of the falsity of non-scientific ways of knowing. This view needs to change.

Intuitive ways of knowing have received considerable attention in diverse disciplines. The findings of many of these studies suggest that intuition lies outside analytical thought and sensory perceptions and within the realm of the unconscious [186]. According to Charles Laughlin the brain due to its unique neurognostic structure and organisation is wired for intuitive modes of knowing. Neurognostic awareness of the "quantum sea" has been a central element in nature for maintaining the status quo between internal homeostasis and adaptation to the external environment [183]. From the 1970's scientists gained a better understanding of the process of brain lateralisation. They found out that cognitive abilities were divided according to right and left hemispheres, with the left hemisphere controlling analytical and numerical thinking while the right controlled spatial awareness and artistic and intuitive modalities [186]. However, more recently scientists have been able to ascertain that the two brain hemispheres are constantly communicating as a integrative continuum [186].

If non-local mind does exist then it may have been acquired through evolution as it may have conferred reproductive and survival advantages to ancestral humans. Certainly, being able to anticipate danger, thereby allowing an individual to survive and reproduce was evolutionary advantageous. Cosmides and Tooby concluded:

*"It may be time to . . . grant human intuition a little more respect than it has recently been receiving. The evolved mechanisms that undergird our intuitions have been subjected to millions of years of field testing against a very rich and complexly structured environment"* [187].

The precarious world of prehistory presented a number of problems which over evolutionary time contoured human fear response. The development of the neocortex added another dimension of fear; unlike other animals, humans could imagine danger even when it was not present [188]. Because the neocortex is linked to more primitive brain structures such as the limbic system, humans can trigger a fear response simply by thinking of a potential threat [188]. This unique fear response in humans may have assisted in the development of intuition in order to bring some semblance of control to an otherwise over-stimulated psyche.

Perhaps, prehistoric shamans developed non-local mind techniques for enhancing group survival such as tracing the behaviours of potential game or alleviating social disruption. In any case, non-local mind may have been an early "form of perception antedating cortical developments in the course of evolution" [189]. Non-local mind techniques may have provided an evolutionary form of information gathering for predicting natural and social phenomena in the ancestral environment. A legacy of this "evolutionary reasoning" are extant divination systems found in many indigenous and traditional cultures. Divination systems often engage in intuitive and creative kinds of thinking in order to access information on everyday events. Schutz's idea of the nature of the life-world is applicable here, since it is based on knowledge stores making sense to others [190, 185]. Such knowledge stores provide templates for social action – a compendium of consciousness of past experiences which inform our lifeworlds [191].

A lot of present day divination is based on relieving individual and social issues and potential problems. We can presume that divination in ancestral times also tended towards problem solving and quelling possible social divisions which may have eroded communal relations. The earliest known form of divination appeared in China over four thousand years ago and involved analysing various animal parts. Among the Jalari of South India divination acts conterminously with stocks of cultural knowledge. In this way, divination acts to explain reasons for social disruptions and illness [192]. In Bangladesh, divination operates as a dialogic process where the client imparts information to the diviner who in turn creatively reworks this information to come to a solution [193]. In this way, divination acts as a symbolic arena of possibility, and a means for diminishing uncertainty.

Among the Kuranko of West Africa divination is not so much concerned with a precise picture of reality but rather a means for gaining a sense of self empowerment in the face of life's indeterminacy [159]. This sense of providing 'voice' in the face of life's difficulties would have been as important to our human ancestors as it is now. Current forms of divination may be categorised into four types: trance, symbol, omen and pattern. Omen divination uses natural signs such as flight of birds or other animal behaviours; trance divination is where the diviner undergoes an altered state of consciousness for communicating with the spirit world; symbol and pattern divinations use various kinds of paraphernalia such as dowsing rods, cards, animal innards, bones, sticks and body features for interpreting the future [186]. Navajo Indians use mediumistic divination where female diviners wash their fore arms and sprinkle corn pollen from their elbows to their right hand palms [186]. After this they close their eyes and visualise communicating with the East Collared Lizard. As the diviner enters into a trance state she begins to move in a jerky fashion emulating the movements of a lizard [186]. After the diviner opens her eyes she may see the afterimage of a physical or symbolic form which indicates what kind of healing ceremony should be performed [186]. This healing nuance of divination is emulated by the Kootenai people of British Columbia who use sweat lodge ceremonies for receiving spiritual healing messages. These intuitive messages are then relayed to other tribal members which are discussed for their meaning [186]. Chagga diviners hold *isale* leaves while beckoning the ancestors for guidance. The divination is based on a dialogic process between client and diviner where the former either confirms or denies the latter's claims. If the diviner's claims are denied they re-scrutinise the leaves [194]. It is because people's lives are enmeshed with material objects, which in turn informs their social relationships, that divination plays a role in reaffirming the nature of Chagga sociality [194]. In contrast, Bamana divination is based on a kind of ethno-mathematics consisting of iterative symbols which are paired and repeated. The process is connected to certain verses (*Odu*) which number 256, that allows for considerable creative manipulation by the diviner [195]. The usage of recursive symbols and verses testifies to the mergence of analytic and creative modes of thinking in Bamana divination.

Considerable scientific research on non-local healing challenges the scientific establishment and supports the possibility of psi. Unfortunately, much of this debate has pitted sceptics and proponents in challenging each other's claims. This has been characterised in many studies which seem to indicate the presence of psi and many studies which do not support this. Who are we to believe? A timely outcome of this debate has been improvements in scientific design methods and more informed analyses when researching psi activity [196].

Many of these studies on distant intentionality were conducted on non-human animals so there was little possibility of a placebo effect. The work of psychologist Bernard Grad during the 1960's deserves mention. In his seminal experiment Grad used 300 mice which were given wounds. The mice were separated into three groups; one hundred mice were the control group and did not receive any treatment, medical students treated one hundred mice, and one hundred mice were treated by a healer. The study was double blind in order to quell any sceptics who may have questioned the experiment's scientific rigour. Following two weeks mice that the healer treated showed significant healing of their wounds compared to the two other groups of mice [197]. Spurred by the results of this seminal experiment, Grad conducted another experiment. This time fourty eight mice had their back injured with identical surgical wounds. The mice were divided into three groups; a control group of 16 mice another 16 mice were put into a cage kept at the same temperature as the third group; the last group of 16 mice were put into a cage that a healer held for 15 minutes twice a day. As in the previous experiment the last group of mice that were held by the healer showed considerably higher healing rates to their surgical wounds "with less than one chance in a thousand that the results were due to chance" [198].

Back in the 1990's an assessment of 30 scientifically rigorous experiments was conducted on distant intentionality. The analysis concluded that it was unlikely that the experiments' results were due to inherent methodological design flaws. While some in the scientific community believe that the results in future studies on distant intentionality may be different due to modifications in scientific designs, it is possible that such studies may reveal a novel way of knowing the world [196]. Another mouse experiment examined the effectiveness of Chinese Qigong treatment on lymphoma growth. Qigong is a Chinese therapeutic method believed to confer various health benefits to practitioners. The technique is based on the Chinese notion that there exists a universal life force called Qi that can be controlled and manipulated through the practice of Qigong. A total of 90 mice were used in the experiment. These were divided into 3 groups; lymphoma cells were injected into 30 mice who were treated with Qigong, 30 mice received no treatment, and 30 mice were given sham treatment. A Qigong healer supposedly emitted Qi to the treated group ten minutes at a time, for five sessions. . A non-healer person who was treating the sham group "imitated the movements of the Qigong healer". On the ninth day 10 of the mice from each group were humanely killed. The preliminary results suggested that the Qigong healer may have impeded the growth of lymphoma cells in the treated mice [199].

In a more recent experiment on healing intention four *Johrei* practitioners directed healing intention for 25-30 minutes at a time towards selected flasks containing human astrocytes in culture. This was emulated during 3 days. During each day 6 flasks were exposed to the *Johrei* healers. Six unexposed flasks containing astrocyte cultures acted as the control group. During each day, 4 healing sessions were conducted [200]. Double blind procedures were rigorously maintained throughout the experiment. The experimental conclusions indicated that repeated healing intention in a given location to treated cell cultures may alter or enhance their growth compared to untreated cell cultures [200].

Another poignant double blind study involved 40 volunteers who had advanced AIDS-associated illness who had received standard medical care. The volunteers were randomly assigned to "receive either distance healing or no healing" [201]. A total of 40 healers from various parts of the United States, who had at least five years experience in distance healing, were used. Healers only knew the first names and photographs of five volunteers. Healing was sent for one hour each day for six days per week. This was repeated for ten weeks. An interesting element was that the healers came from various traditional religious and shamanic backgrounds. At six months results indicated that those volunteers who had received distance healing had lower rates of AIDS-associated illnesses and had lower severity of existing symptoms [201].

Since the 1990's there have been new scientific theories which have attempted to explain non-local mind. One of the most novel explanations has been developed by the prodigious scholar Ervin Laszlo. Over many years Laszlo formulated a theory called quantum vacuum field, which later was re-formulated as the Akhashic field. The word Akhashic derives from Hindu notion of the Akhashic record that contains the memory of everything that has existed in the universe. Similarly, the Akhashic field is explained as a void, vacuum or containing cosmic space [202]. This immanent field is produced by wavelets that spread across space-time recording the entire cosmic memory [203]. All existence comes into being and is recorded by the Akashic field since – this infinite field contains the entire universal experience of everything that has ever existed [204]. The twentieth century genius Nikola Tesla in a 1907 paper had stated that there exists an immanent cosmic field through which everything can be referred to it [202].

Related to the Akhashic field is the notion of the Metaverse. Various scientists have considered the Metaverse as a vaster probably infinite cosmos, that preceded the existence of the known universe, and that will continue to be present when the current universe ceases to exist. Various astrophysicists have recently discussed the possibility of multiple universes of varying physical constants. One calculation estimates that there may exist 101010,000,000 parallel universes [205]. The Big Bang is considered as a quantum event that expanded rapidly. During this period of inflation quantum fluctuations were 'set' as "classical perturbations" eventually becoming distinct universes [205]. Laszlo argues that our universe takes its origin in the Metaverse that, possibly spawned previous universes, and in the future will inform new universes. The Metaverse is the point of origination of local universes, attesting to their order and "remarkable coherence" in general [202]. As Laszlo explains:

*"The same way as the genetic code of human parents informs the fetus…the Metaverse informed the Big Bang, the otherwise inexplicably precise explosion that gave rise to this astonishingly coherent life-bearing universe. It also gave, gives, and will give birth to other universes, producing periodic universe-creating explosions…this kind of evolution we observe in our own universe got under way, and will get under way, time after time"* [202].

The Metaverse is the infinite depository of information-templates from pre-existing universes [206]. Thus, our universe is one of illimitable possible previous universes of the Metaverse. The Metaverse is forever immersed in its own kaleidoscopic play – an endless cycle of crunches and rebirths [206]. The idea of a universe enmeshed in an endless cycle of births and destructions was first conceived of by Hinduism thousands of years ago. Hindu thought professes that this universe is a product of the dreaming god Vishnu, and exists for one hundred Brahmanic years. It is here that the Hindu play on numbers becomes really impressive. A day of Brahma is an equivalent of 4,320 million years. One cycle of Brahma equals 1022, after which the universe is dissolved, only to begin again in another cycle.

A third important concept of Laszlo's is coherence which refers to the interconnectedness of everything from the largest to smallest structures of matter. Coherence is a necessary requirement of life to exist [202]. Put in another way, coherence is the integument in the fabric of the cosmos. In evolutionary life coherence is based on the fact that all life on Earth is related by virtue of their DNA structure. In human terms, coherence is characterized by the many variations of sociality found throughout the world. The link here between non-local mind is that certain individuals may be able to tap into the Akhashic field and access its limitless storehouse of knowledge. Akhasic field may be a source creating new kinds of coherence between non-human and human worlds. The Akhashic field may have profound implications for the coherence of life on earth and the planet's connection with the cosmos.

## THE SHAPESHIFTERS

A fascinating element of shamanism involves the belief and practice of shapeshifting which is defined as a shaman transforming into a non-human animal, and also includes behaving like one. The belief of shapeshifting stems from prehistoric times in various parts of the world. Prehistoric paintings in Lascaux (France) [207], Altamira (Spain), Africa, [208, 209],] the Americas and Australia show theriomorphic figures which have been a source of speculation by scholars. Outstanding examples of prehistoric therianthropic figures are "the Sorcerer" and the "Lion Man of Hohlenstein". The former refers to an enigmatic cave painting (13,000 BP) in Trois-Frères, Ariège, France, which depicts a humanoid figure with horns and tail and is probably a representation of a shaman.

The second figure which was found at Hohlenstein mountain in southwest Germany, is the oldest known therianthropic statue (30,000 BP) carved from mammoth ivory. The statue depicts a human figure with a lion's head or a mask worn by a shaman. The therianthropic legacy of ancestral humans was passed down to classical civilisations whose religions had an abundance of humanimal deities and myths. For example, in Hinduism the god Vishnu is believed to have had ten incarnations known as the *dashavatara*. Some of these were in therianthropic forms such as *Matsya* (fishman), *Kurma* (tortiseman), *Varaha* (boarman), and *Narasimha* (*lionman*). Each of these incarnations came at a time when there was great evil in the world. Their role was to save humanity. The noted biologist J.B.S. Haldane developed an interesting theory of the *dashavatara* by proposing that each incarnation represented a developmental evolutionary period. For example, *Matsya* represents the advent of vertebrates while *Kurma* represents the dawn of reptiles. The ninth incarnation is the historical Buddha who represents humanity during the Axial age (800-200 BCE), a time when the classical civilisations had synthesised their moral teachings.

Widespread belief in shapeshifting is probably a remnant of a time when ancestral humans had co-evolved with the non-human world. Ancestral humans would have been highly observant of the behaviours of many animals which was essential for survival during the Pleistocene period. Early shamans carefully scrutinised the non-human world to develop myths and ritual practices. Shamans had probably observed insect metamorphosis where their bodies undergo total transformation. Metamorphosis would have been considered as a magical transformation akin to rebirth. Being acutely aware of their mortality ancestral humans found in metamorphosis a way of imagining life beyond death. Shamanic imitation of metamorphs further provided insight into the nature of non-humans. In extant shamanic cultures shamans behave in animal ways in order to embody their ways of thinking and behaviours. This human penchant for mimicry of the non-human world probably led to the advent of totemism. Mimicking non-humans may also have informed music by copying the sounds of birds. In this way, music was not a unique creation of humans but rather a rearrangement of bird songs.

Western fascination with shapeshifting has been predominantly characterised by werewolves (lycanthropes). Werewolf superstitions were known in Mesopotamian, Indian, Greek and Roman lore. According to the Greek historian Herodotus, the ancient Neurians changed themselves into wolves [210]. Among the Romans the werewolf was known as *versipellus* or 'skin-changer' [210]. One theory purports that werewolf beliefs during the Middle Ages had been informed by earlier Norse warriors known as 'berserker' who had fought in a frenzy during battle wearing wolf pelts. The wolf clad berserker's battle fury became synonymous with the atavistic werewolf. Clothed in wolf and bear skins, the berserker indulged in maniacal displays of violence, roaming the evenings like enraged predators - the stuff of nightmares [210, 211]. It has even been argued that lycanthropy was an atavistic desire to return to a state of early hominins who had been vegetarians [210]. While it is problematic to suggest that lycanthropic traditions are a kind of subconscious retrieval of former animals, they do point to human fascination with becoming an 'animal other'.

Hollywood has been an avid proponent of the lycanthrope myth producing many movies on this theme. The most recent cinema addition has been the popular Underworld trilogy where werewolves (lycans) are pitted against vampires in a clandestine battle over centuries. In the movies the more developed lycans have the ability to shapeshift at will while the oldest lycans can never retrieve their human form. The portrayal of the lycans accords with werewolf stereotypes as being bestial and lacking any ability to think beyond their basic instinctual drives. The 1994 movie *Wolf* starring Jack Nicholson depicts him as a businessman who is bitten by a wolf. After this he undergoes a peculiar transformation. His senses of smell, hearing and strength become out of human range. His olfaction becomes so acute that he smells the odour of his wife's secret lover on her clothes.

But why so much interest in werewolves? Possibly it could be because wolves were the first animals which our species domesticated. DNA from 1500 dogs was recently collected and analysed [212]. It was found that present day dogs originated from a few hundred wolves between 10,000 and 15,000 years ago in southern China [212]. This Chinese human/wolf connection was characterised by P'an Hu, a shapeshifting dog-headed man who had married a Chinese princess and fathered a human race [213]. It has been surmised that this wolf cohort ventured close to humans in order to access food scraps. The pickings were so good that these wolves stuck around and eventually became domesticated. Dogs are probably the only non-human animal species that took an active part in their own domestication. During the 1950's the Russian scientist Dimitri Belyaev conducted genetic experiments on the silver fox in order to understand how wolves became domesticated. These experiments continued right up until the 1980's allowing sufficient time to answer this question. Several interesting events happened during the experiments. As foxes were selectively bred for tameness they started to look and behave differently. Fox puppies competed for human attention, whined and wagged their tails [214]. Moreover, adult foxes retained juvenile 'dog' features such as floppy ears and raised tails, while adult males had smaller and 'feminised' heads, unlike wild foxes [214]. Domesticated fox pups also opened their eyes a day earlier than their wild kin, and took longer to have a fear response (week nine *versus* week six in wild foxes) [214]. Other physiological changes included less adrenaline in the blood and higher levels of serotonin and its metabolite 5-oxyindolacetic acid which made the foxes less aggressive and more docile. Even the reproduction cycles of these domesticated foxes became more 'doglike' where females had out of season mating not found in wild foxes [214].

The identification of a shaman with animal powers reaches its zenith in the putative ability for him/her to turn into a non-human animal. There are numerous stories testifying to shamanic shapeshifting proclivity. Some of the most fascinating shapeshifting stories come from Siberia during the Soviet era. One story involves shaman called Parilop of the Srednaia Kolyma region of Russia who was known to have unusual precognitive and healing abilities. He is believed to have foreseen the burning of an electric power station a week before the event had occurred. Parilop had woodgrouse as his spirit helper. One day Parilop had decided to visit his friend Kharkha who lived in the village Khatingnaakh as a woodgrouse. When Kharkha saw the bird looking at him he tried to shoot it, however, the bird escaped. After a few days, Parilop went to Kharka's house and said, "You nearly shot me the other day before yesterday. Why did you try to do this?" [215]. A similar story is based on the testimony of Vasily A. Kudrin who had been a Communist League member. Kudrin narrated the story of the Kolyma shaman called Gul'aev who had died in 1965. The shaman was searching for a successor, however, Kudrin turned him down.

*"Suddenly I saw a raven swoop down and scare the horses. I calmed them down and went on. The next day, Gul'aev found me in the street and said ..."I've been looking for you and finally found you". I was petrified. He had seen me react with the horses when he was in the form of a raven and wanted me to then become his apprentice"* [215].

Among the Sakha shamans spirit helpers in the forms of wild and domesticated non-human animals are used to mediate his powers. Spirit helpers are often ravens or other birds. During séances Sakha shamans engage with avian spirit helpers. These spirit helpers assist the shaman in reaching cosmic terrains where he/she can retrieve lost souls or find sickness spirits. Powerful shamans may have as many as 47 spirit helpers [215]. In Sakha shamanism the children of the sky gods Ary Darkhan may take the form of an eagle or raven and mentor the soul in various shamanic wiles and skills [216]. A unique feature of Sakha shamanism is the belief that shamans possess a mother soul which can travel to the spirit world where it is transformed to a iiƏ-kyyl (mother beast) [215]. The iiƏ-kyyl is intimately tied to the shaman for when it dies the shaman also dies [215]. A shaman's iiƏ-kyyl can fight with other shamans for magical supremacy. In one story relating to the aforementioned shaman Parilop, he is alleged to have taken revenge on an Evenk shaman by becoming a wolf and killing the latter's reindeer [215]. The shapeshifting ability of a Sakha shaman is tied to the kinds of spirit helpers he/she possesses. Typically more powerful shamans may have an eagle and bear spirit helpers while weaker shamans may have a dog or wolf. Attaining the help of a spirit eagle is difficult as attested in the following:

*"He stood, and bowing to all people present said, "Hello! Why have you called me here? Why have you bothered me? With what will you reward me? My sharp fast eyes are piercing...my eagle talons capture and pin you, I will squeeze anyone who clutches my tail! Why have you asked me here?"* [215]

Implicit in the afore-mentioned narratives is a shaman's ability to reorientate to the contours of the non-human world in ways that are oblivious to monophasic westerners. Shapeshifting exemplifies a kind of psychic acuity to the environment. Shapeshifting begins at the onset of life. Humans are born shapeshifters as the process of becoming a human being entails a lengthy period of socialisation requiring neonates to mimic others. Enculturation is mandatory in order to achieve the transformation from human animal to human being. In feral children where this process is absent we find that their senses and behaviours mimic their animal host. Even when captured such children can acquire human language but never fully master it. It is as if the 'innate animalness' of our bodies takes precedence over our cultural selves.

In shamanic cultures mimicking non-human others plays an important part in socialising youngsters to requirements of their culture. For instance, Aboriginal people of Australia frequently use ceremonies, songs and art to commune with the Dreaming ancestors. Aboriginal relationship with the land is primarily informed *via* their concept called Dreaming. The Dreaming refers to a primordial period when ancestral beings created the world. These beings created the landmarks of the earth as well as animal and plant species. They left their tracks in trails, mountains, hills, valleys, rivers, lakes, caves and other natural features in the land. When the earth had been formed, the ancestral spirits returned back to the earth or went to the stars. While the Dreaming discusses the creation period of the earth it still informs the present.

Individuals in shamanic societies are enculturated to be hyper-sensitive to the movements of creatures and changes in weather patterns and the landscape. However, it is not enough to know non-human behaviour. For the shaman he/she must become the other if they are to know it. Their bodies must become porous in order to become a receptacle for the other.

A key element in such kinetic invocations of another animal is the magician's ability to dream himself in the wild physicality of that other. Allowing his senses to heighten and intensify as he becomes possessed by the carnal intelligence of the creature [217].

In his poignant work *Becoming Human*, David Abram discusses how his shaman friend called Sonam had tried to teach him of mimicking certain animals. Sonam was highly accomplished in animal mimicry, able to utter the squeaks, gutturals and croaks of ravens with such precision that they would swerve towards him [217]. Sonam was also adroit in copying raven postures and ways of walking.

Abram explains that to behave as the other is the most visceral method to feeling oneself into the body of an animal. An important element in this kind of "kinetic invocation" is allowing one's senses to intensify and heighten during the act of mimicking the other [217]. The shaman must feel himself/herself that they are engaged in a metamorphosis. This necessitates a rearrangement of a shaman's senses and thinking so that they merely not fake the other but rather become the other. Under Sonam's careful guidance Abram spent arduous weeks recalibrating his senses in order to become more conscious of the animal world. During one exercise Abram had observed a raven nearby. As Abram's gaze shifted from the raven's eyes to its body it would fly away [217]. While it seems hardly possible that the raven could have detected a shift in Abram's attention, this is exactly what had happened several times. Sonam had instructed Abram to use his senses in a different way in order to fathom his visceral domain. The telos of Abram's practices came one day while he was walking towards his hut. On the way he came across a raven picking at a carcass. Abram notes that as the raven pecked at the carcass he could feel a sensation in his neck. As the bird loosened bits of meat from the carcass Abram felt his chest becoming increasingly weightless. It was as though the bird's pulling apart the carcass found its analogue in Abram's body [217]. Abram had undergone the shamanic experience of being intimate with the other.

Abram's experiences of becoming the other are mirrored among Sufis in North India whom AS had engaged with during anthropological fieldwork in the mid 1990s. Like shamans, Sufis are deeply involved with communion with the spirit world which is the source of their putative mystical powers such as precognition. Sufis spend much of their time engaged in their mystical practices which include frequent fasting (*roza*), sequestering (*khalwat*), verbal and non-verbal mantric exercises (*wazifa*), and prayer (*dua*). All Sufis engage in some kind of altered state of consciousness in order to gain spiritual power. The intense nature of Sufi mystical practices seems to endow them with heightened sensitivity; their senses seemingly become more acute enabling to engage with the environment in different ways. For example, Sufi practice of *wazifa* involves reciting non-verbally one of the divine names of God several thousand times. The practice co-ordinates the mind reciting the divine name with the breath. The practice emphasises feeling the divine name circulating within the body's viscera. One Sufi claimed that the divine name circulates throughout the body cleansing the heart. Implicit in Sufi narratives is the development of a sensuous awareness of their bodies in relation to the sacred other [218]. This sensuous awareness is evident in Sufis' engagement in holy places. Literally, Sufis see, feel, hear and taste the presence of the saints during their peregrinations to holy shrines; the cadence of their supplications, prayers and sacred chants invoke the saints' spiritual emanation (*barkat*) into themselves [218]. Sufis constantly engage their sensory perceptions in order to apperceive the nature of the landscape. In one of AS's initial meetings with a Sufi Ahmad Shah, he took him throughout the famous Nizamuddin shrine and the surrounding village called the *basti*. Ahmad Shah possessed intimate knowledge of the shrine and *basti*. He would tell stories of certain places and whether there were good or bad spirits there. He was apparently able to apperceive the nature of a certain place by the spiritual residue emanating from there.

Another interesting element of Sufis is their seeming mimicry of spirit beings referred to as *jinn*. In Muslim lore the *jinn* refer to a race of invisible spirit beings which were created by God before the advent of humankind. The *jinn* are believed to possess all manner of supernatural powers and are ambiguous by nature. The *jinn* are said to roam the world at night and can possess people whom they are attracted to. They also like to habituate dark and isolated places so that people cannot disturb them. So how do Sufis mimic the jinn? Firstly, the nature of the *jinn* is emulated by the ambiguity of Sufi behaviours. Some Sufis are called *jalali*, (referring to unusual power) due to their unusual precognitive and healing powers. However, being *jalali* is also associated with wrath, awe and indeterminacy. Like the *jinn*, Sufi emotions are fluid, changing from kindness to anger in an instant. This is because the power inside their bodies cannot be contained and must be regularly unleashed. A Sufi's *jalali* nature makes him desire loneliness for he cannot bear the mendacity of the world. A Sufi once told AS that while peaceful Sufis may admonish others for being dishonest a *jalali* Sufi would be inclined to "fix him/her up". This could mean sometimes partaking in sorcery. Furthermore, Sufi mimicking of the *jinn* is evident by Sufis habituating lonely places where *jinn* are supposed to reside. While ordinary people fear the *jinn*, *jalali* Sufis often invoke the *muwakil* (spirit familiar) in wells, caves, cemeteries and other lonely places to assist them in their mystical practices. Older Sufis may have several *muwakil* helping them. *Jalali* Sufis must go through arduous magical practices in order to 'catch' a *muwakil*. Even then, it is said that a *muwakil* cannot be fully controlled and can destroy a less powerful Sufi. The *jinn's* affinity for the night is also reflected in Sufis' engagement in their mystical practices during this time. The night is viewed as a time when mysterious powers can be invoked. While Sufis engage in healing (*ruhaniyyat*) during the day, sorcery (*jadu*) is committed during night. This separation between the day and night is reflected in Kuranko (a group in Sierra Leone) cosmology where daytime symbolises public life. In contrast, night denotes wild and inscrutable power [219]. AS remembers how his Sufi friend Shams would sometimes take him to visit saints' shrines during the night. As he approached these places he would enter into an altered state of consciousness where he would seemingly enter in conversation with the spirit world. Shams would praise the saints by performing special prayers, lighting incense and even eating dirt around a saint's grave. He said that this was *tariqah* – A Sufi's mystical path. One night Shams performed a series of mystical gestures at the shrine of Inayat Khan. During his performance he entered into an altered state of consciousness, evinced by his laughter and dancing around the saint's tomb. His ebullient manner reflected a Sufi's ability to act in ways contrary to social norms.

Like the *jinn*, a Sufi's motility is governed by stealth, usually walking alone and during the night. His thin body, an artefact of regular fasting may also be viewed as a kind of invisibility. During the day, Sufis usually sit in a foetal position which is conducive to meditative states, sometimes covering their faces and upper bodies in a shawl so as to make them appear less visible.

Like the *jinn*, Sufis are noted to cross boundaries. AS remembers how one Sufi would sit in the woman's part of a mosque at the chagrin of male Muslims. By seeking to become the 'other' Sufis had reinvented themselves in accordance with the spirit world. This somatic mimicking of otherness was a means of establishing anontological connection with the other; a kind of virtual reality for 'feeling' the other [220]. Lastly, Sufi bodies are invariably thin from frequent fasting; in this case being thin inclines towards being less visible.

The human body is given to shapeshifting by virtue of its external senses which draw and entangle the body with the environment [217]. Rather than being static as believed in western medicine, the body is an indeterminate plenum which ebbs and flows according to diurnal and seasonal cycles, always in the act of being transformed. Indeed, the body's metamorphic capacities come to fruition every seven years, the amount of time it takes for every cell of the body to be renewed. The implication of the body's renewal with the number seven offers some tantalising insights. Firstly, the number seven is connected with the polyphasic world of prophecies, dreams, and the mantric arts [221]. In pre-Homeric Greece, prior to the rise of priesthoods, Hellenic women acted as oracles where they communicated with the earth and underworld spirits. Moreover, the number seven represents the stages of human cycles and reflects the peregrinations of the seven planets known to the ancients. The number seven also governs the seven moon cycles, the seven-note scale in music and the seven lights of the Great Bear that inform time [221]. The affinities linked to seven suggested here deal with cosmic and terrestrial cycles of transformation.

Indeed, current science actually attests to the shapeshifting powers of nature. The process called symbiogenesis requires organisms acquiring other organisms, not just their traits [222]. The act of symbiogenesis also generates novelty, propelling the emergence of new life forms. The tendency for life to co-evolve and co-exist is inherent throughout nature and perhaps the entire cosmos. Prokaryotic life forms were the first shapeshifters over 3.5 billion years ago. The first of shapeshfiting steps was the mysterious development of life from amino acids and elements, an incredibly improbable process. The next shapeshifting feat came when different cells started to merge with other cells in order to improve their chances for survival in an increasingly hostile environment. These primordial mergers assisted in the development of Eukaryotes 2 billion years ago, each heralding more complex cellular organisation such as a nucleus and mitochondria. The third shapeshifting development arose in response to increasing levels of atmospheric oxygen during the Proterozoic Era, necessitating the development of respiration which is inherent to all aerobic based organisms. A third creative development was the movement of life onto land by plants (425 million years ago) and animals (395 million years ago) which demanded a new kinds of bodies in order to deal with the challenges of gravity and an arid landscape. Each of these achievements generated novel ways for adapting to ever changing environments.

This constantly evolving facet of life is central to shamanic cultures. Life is both transient and contingent on patterns of connection. These patterns engage human bodies to the contours of the landscape and non-human animals, leading to an ecology of cyclicity and regeneration. This notion of circularity and change is depicted in the shapeshifting theory of Kung San people of the Kalahari. In Kung San thought shapeshifting is referred to as *thuru*. It is *thuru* that enables the Big God to create life [223]. Another important concept is *G//aoan* which refers to god, devil and ancestral spirits. *G//aoan* also creates lesser forms such as good and bad spirits and characterises "the primal process of transformation" that underpins the circular logic of the Kung San. To practice *thuru* is to embody the inter-changeability of existence [223].

A modern twist of shapeshifting comes in the way of transhumanism which is based on future humans being augmented *via* various bio and nanotechnologies. In transhumanist way of thinking a transgenic new human reaches beyond our biological limitations so that it may become a species of humans different from *Homo sapiens*. More than a hundred years ago H. G. Wells introduced the concept of the transgenic posthuman (new generation of augmented humans) in his novel, *The Island of Doctor Moreau* (1896). The book describes efforts of a deranged scientist to use animals to construct a 'perfect' human. Moreau produces a series of humanimals that are dystopic individuals who cannot remain human and tend to revert to their animal substrates. It follows that boundaries between animals and humans should not be crossed.

For over a billion years evolution has perfected ways of improving living organisms. Algorithms created by humans aim to improve efficiency of evolution to levels not achieved before. Nick Bostrom, a philosopher, thinks that advances in nanotechnology and computational neuroscience may lead to a novel understanding of the human mind. A transhumanist Ray Kurzweil concurs with Bostrom that artificial intelligence may match human intelligence by 2020. This will make Moore's Law redundant. Moore's law describes the trend in computational hardware and includes the driving force behind technological and sociological levels of society. According to Moore's law there will be a tenfold computer microprocessing increase within the next decade. After 2020 a combination of nanotechnology and quantum computers, predicts Kurtzweil, will make computational singularity possible meaning that the exponential growth of "computer intelligence" will reach infinity. In 2060 computers will be able to simulate power of one billion human brains while by 2099 computing power will exceed billion times brain power of all people in the world [224]. Eric Drexler offers a different perspective. He argues that a sugar-cube-sized computer can run at 1027 operations per second [225]. Seth Lloyd provides an estimate of an even greater computing power [226].

A brain genome study currently underway hopes to improve our understanding of biological neural networks. This understanding may lead to a completely new approach to mind sciences that will enhance our cognitive potential [227]. Advances in neurosciences may be coupled with advances in artificial computer intelligence leading to new kinds of solutions based on combinations of biological brains with digital devices. It may be possible, according to Bostrom, to use brain-machine interfaces to upload human minds into the cyberspace. This may require detailed analysis of brain cells in order to run an emulation of their interconnections on a computer [228]. While repairing of biological brains in order to preserve minds is not a viable option at present, mind uploading seems viable. Were it so, people would be able to have simulated copies of their minds that could survive longer than their biological bodies [229].

Like Bostrom *et al.*, the physicist Frank Tipler states that post humans will be compelled to gain mastery over matter on atomic and molecular levels, and that such mastery will lead to an exponential increase in human information. Tipler's solution concurs with the post-human thesis that humans must transform their bodies, preferably in the virtual world of cyberspace. According to Tipler, machines will increase their computational capacity to feasibly upload human minds by 2100. Such mind uploads would then be placed in small spaceships in order to colonise the Milky Way galaxy which will take approximately one million years [230]. From there, space probes will reach and colonise the distant Virgo cluster in one hundred million years time [230].

Allhoff points out that the boundary between machine and a human will become less clear when technological advances allow use of prosthetic neural implants [231]. Then a new definition of a human being will be needed. Furthermore Susan Greenfield, a neuroscientist, maintains that at the end of the 21$^{st}$ century the way people conceive their bodies will be changed. In her understanding human brains will be connected to computers enabling insight into a number of virtual worlds. Such virtual worlds will be preferable to the actual reality leading to people paying less and less attention to the real world around them [232]. However, Greenfield's views are not shared by transhumanists.

One may ask at this point, what does post-human existence offer? This is a pertinent question. Post-humanism offers future humans a virtual world where they may freely be able to alter their appearance without being stuck in a physical body. Post-humans may also be able to experience and enjoy various virtual utopia, and savour every kind of culinary and sensual pleasure, and be in contact with other virtual beings [233]. In this virtual playground new dimensions of sexuality may also be experienced without the problems of unwanted pregnancy and sexually transmitted diseases [234]. Although future humans will be encouraged to shed their biological bodies this will not mean an end to physicality and sexuality.

A different transhumanist vision is presented by Paul Ehrlich who considers that humans created from splicing between human and non-human animal DNA (transgenic humans) due to their non-human sensory enhancements may be more attuned to the environment [88]. Were people able to see ultraviolet light, Ehrlich says, they would be able to perceive the extent of the depletion of the ozone layer [88]. By the same token, if people had greater olfactory perception they would be able to sense chemical pollution of the air.

## THE ALCHEMICAL WORLD OF DREAMS

For millennia, both shamanic and traditional societies have been using dreams as a source of knowledge about self and the world. Western societies have treated dreams as separate "things" distinguishable from waking consciousness. This separation sharply contrasts with indigenous peoples such as the Ojibwa of

Canada who do not make a distinction between dreaming and waking states [235]. For the Ratamuri of North America dreams are real events and are a way of communicating with the deities [236]. Similarly, the Tenne of West Africa [237] and the Ilahita Arapesh of New Guinea [238] dreams offer an opportunity to communicate with the ancestors. For the Gebusi of New Guinea sleep is a time when the soul departs the body and travels to the spirit world [239].

Among Native American cultures dreams are often similar to vision quests whereby an individual seeks guidance from the spirit world. Among the Menomini people individuals will sometimes fast for several days in order to have a visionary dream, such as the 'Tall Man from the East', who is a culturally identifiable symbol. Such culturally sought after dreams were found among ancient Mediterranean peoples who would visit temples to the healing god Asclepius [240]. In medieval Japan, samurai warriors would seek the guidance of spirits (*kami*) at specific shrines. After his defeat by the famed swordsman Musashi, Gonosuke performed purifying rituals at a Shinto shrine for 37 days. During this time he had a dream where a divine being told him how to perfect his martial skills. After the dream Gonosuke developed a formidable art of the short staff.

During fieldwork in North India AS also analysed Sufi dream telling. Indian Sufis have existed in India for at least one thousand years and like their Hindu counterparts – the *sadhus* have developed an elegant repertoire of ascetic and mystical practices over the centuries. Among these practices are dream visions and dream telling. The dream world is especially auspicious for Sufis as it is a source of communicating with beneficent spirit beings such as saints. AS was told that Sufis may be initiated in the inner mysteries by the saints which are sometimes referred to as *pir-ghaib* (the invisible teacher). The *pir-ghaib* is synonymous to the spirit guides of shamans where they confer special powers. As in other cultures, dreams are a source of creative insight and enable Sufis to engage in other ways of knowing. Sufis categorise dreams according to instructional dreams by spirit beings (*basharat*), and precognitive dreams (*ruhi*). Many dreams are apparently concerned with spiritual struggles facing Sufis where they learn how to temper his lower self (*nafs*) [241]. An interesting element of Sufi dream culture involves dream interpretation. Early in AS fieldwork it was found out that Sufis were interested in AS dreams and were more than eager to interpret them. For months AS became privy to the mysterious world of Sufi dream symbolism.

Sufis were adroit in providing an insightful dream exegesis which involved analysing a complexity of symbols. Sufi dream analysis was based on both Indian Muslim folklore and on their own mystical experiences. In one dream two black dogs held AS's two arms in their jaws. A tall old man by his side looked at him soberly. The next day when AS told the dream to one old Sufi he told him that he was the tall man AS had dreamt about and had come to guide him from the clutches of his lower self. In another dream AS was surrounded by lions which did not attack him. This dream was interpreted by another Sufi as being auspicious [241]. In Sufi dream lore, a lion symbolises Ali ibn Talib, who was a cousin to the Prophet Muhammad and one of the righteous caliphs in Islam. Ali is also viewed as a spiritual custodian of all Sufi orders and is conferred with great mystical powers. An important element of Sufi dream analysis is that it provides an opportunity for Sufis to exercise their creative insight by playing with the psyche. For example, in the afore-mentioned dream where the Sufi had psychically entered AS's dream world, is synonymous with that among the Tibetan *Yolmo* shamans who are believed "to transcend the boundaries between Yolmo bodies and thus breach the borders between tacit and apparent realms of experience" [242].

A fascinating element of dreams is their ability to evoke creativity. Shamans have always known that the dream world is fecund with creative possibilities. In better known circles the musicians Mozart, Tartini, Wagner and Beethoven credited some of their works as being inspired by dreams. The scientist Frederick Kekulé is credited to have discovered the structure of the benzene molecule ($C_6H_6$) in a dream where he saw many snakes conjoining to form a whirling circle [243]. Another strange story is based on the archaeologist Hermann Hilprecht who in 1893 solved a famous riddle of an Assyrian inscription. In his dream Hilprecht was confronted by an ancient Assyrian priest who told him that he had incorrectly classified fragments and that the fragments were connected with the god *Ninib*. Upon waking Hilprecht was able to verify with complete accuracy everything that the priest had told him [243]. The Indian mathematician Srinivasa Ramanujan stated that the Hindu goddess Namakkal appeared in his dreams and presented him with mathematical formulae. Ramanujan had no formal schooling in mathematics.

*"While asleep I had an unusual experience. There was a red screen formed by flowing blood as it were. I was observing it. Suddenly a hand began to write on the screen. I became all attention. That hand wrote a number of results in elliptic integrals. They stuck to my mind. As soon as I woke up, I committed them to writing"* [244].

The renowned twentieth century physicist Niels Bohr whose discoveries led to formulating the structure of atoms and quantum theory, developed his ideas of these through dreams. In one dream he saw marked lanes on a horse racing track. He then constructed by correspondence the electron orbits moving around atomic nuclei. This dream inspired the creation of quantum theory [245].

The Mabube weavers among the Tukular are renowned for using dreams to create their weaving designs. In addition, dreams are a source for transmitting weaving knowledge and expertise [246]. Similarly, for the Bardi people of Australia dreams are revelatory; in their dreams Bardi ancestors disclose new kinds of ceremonies, designs and sacred songs [241]. Recently, in a native title claim in the Federal Court of Australia a Ngarinyin man told the court that he had acquired certain sacred songs in dreams.

So I was longing on my bed, on my blankets, I went to sleep. My grandfather came along and told me, "Wake up my grandson. I got something to give you". He said. "I'll give you the song: you must remember this". That's how I got the song. He showed me three songs [241].

In native American Mohave culture various kinds of knowledge and skills are acquired through dreams. For the Mohave learning a kind of knowledge is considered to be ineffective unless it is dreamed. Healing songs, are one instance, of having been verified through dreams [248]. The Lelet of New Guinea use dreams for verifying truth *via* communicating with human and non-human spirits. Lelet have reported that their dreams have enabled them to create artistic genre such as songs and mask [249]. As the Lelet converted to Pentecostal Christians their dreams became a source of supernatural powers confirmed by the Holy Spirit [249]. The transformation of Lelet dreaming is indicative of how dreams assist in providing meaning in the face of social change. Unsurprisingly, dreams often occur when people are struggling to make sense of what is happening to them [250]. For example, native American prophets were instructed by their dreams to reveal new religious teachings in response to European settlement [251].

Dream responses to certain life crises lead us to the evolution of dreams. Recent research has unearthed that dreaming probably evolved in humans in order to simulate responses found in waking consciousness such as threatening events and circumstances [252]. The ancestral environment in which humans evolved was highly precarious so this theory makes sense. Dreams may also have functioned as a rehearsal to play out inimical episodes as a way of engendering better responses to real life threats. This dream legacy has been used by shamans since ancestral times such as dream divination where a shaman seeks to find a solution to a patient's problem. The dream rehearsal theory to real life threats is seemingly corroborated by dream data reports by Hall and Van de Castle. In their study of over 500 dreams they found that over 80% of dreams consisted of negative emotions while under 20% of dreams consisted of positive emotions [253]. The evidence suggests a disproportionate bias to threatening elements in dreams.

While evolution may have informed the emergence and function of dreams, the things which humans dream seem to be gendered. That is, men and women seem to have sex specific dreams that reflect their psycho-biological makeup. In a study conducted by University of the West of England researcher, Jennie Parker 100 females and 93 males (aged 18-25) were examined to find out what kind of dreams they had. Women reported to have more nightmares than men, and that these nightmares were more emotionally intense than in their male counterparts. Furthermore, women's dreams had more family members, were less aggressive and had more self-negativity and misfortune. In contrast, men had more dreams where they engaged in sexual intercourse and had more physical aggression [254]. This gendered propensity in dreaming in women may reflect their neurological make up which makes them worry and ruminate more than men. For example, women have a larger and more active anterior cingulate gyrus which may give them a greater capacity to feel empathy, as well as, the tendency to worry more.

On the other hand, men have larger amygdala embedded in the limbic system which process fight and flight responses [255].

In another study it was found that 98% of subjects (1,140 American college students) had reported at least seven types of dream enacting behaviour. In other words, 93% of students who had dreamt of fear woke up fearful, while 72% of students who had happy dreams woke up smiling [256]. As in the previous study women had reported more fear, speaking and smiling in dreams while men had more dreams where they were sexually aroused. The findings suggest that dreams may have a neurological basis in women and men. Dreams may act as a bio-feedback system where dreamers play out emotions which may inform waking consciousness.

Dreams may also have been selected by natural selection as a way of alleviating stress. During sleep the stress neurotransmitter norepinephrine is diminished. Dreams, therefore, provide a unique neuro-chemical therapy that calms emotional responses and alleviates stressors [257]. Recently, dream researchers have agreed that dreaming occurs during REM, NREM and sleep onset [258]. It follows that, sleep states should not be viewed as separate phases but rather as a continuum [247]. An interesting element of dreaming is that the dream self seems to be more fluid than the waking self. This kind of independence from a conscious self has been referred to as the autonomous imagination [259]. The autonomous imagination endows the dream self with unalloyed freedom to interact with the dream world, and that this process may be responsible for the high level of auto-creativity in dreams [260]. The high level of creativity and problem solving in dreams was the focus of a 1970s study in which 200 American college students were asked if they had during sleep reached to a conclusion to a specific problem. One third of students stated that they had found a solution [261]. Experts from the University of California have figured out why sleep is good for problem solving. REM sleep seems to produce solutions by stimulating associative brain networks to piece together nonrelated concepts [262]. In a Remote Associates Test (RAT) people were given "multiple groups of three words" (*i.e.*; heart, cookie, sixteen). They were then asked to propose a fourth associative word (in this case the word 'sweet'). Participants were tested twice a day: in the morning and afternoon. The afternoon test took place after a nap with either NREM or REM sleep. Participants were also placed according to REM, NREM and quiet rest groups. It was found that approximately 40% of participants in the REM sleep group had improved over their morning tests [262].

In another current study conducted to test the relationship between dreams and creativity, psychologist David Foulkes found that children who had dreamt the least often had lower imaginative ability, even though they had normal verbal and memory skills [263]. On this note, the imaginative ability in childhood is retained into adulthood in many shamanic cultures. This could be one reason why shamanic cultures tend to experiment more with altered states of consciousness. Modern dream research findings act as a warning about current sleeping patterns. Modern humans are sleeping less and less due to the fast pace lifestyle and large scale lighting of our societies, thereby, depriving ourselves of the imaginative trophic factor of dreams. Modern children are also sleeping 1 hour and 15 minutes less than children who lived in 1897 [264]. In contrast, indigenous and traditional cultures tend to sleep more. The distinctive features of sleep in western industrialised societies which include sleeping in isolation, in long bouts, and according to schedules based on mechanical time regimes may not be optimal for inducing dream states [239].

## BEING LIKE GREEKS

One may ask whether in this age of science, of 'magical' technology there is a place for polyphasic modes of consciousness? Probably, this question is best answered by an examination of the ancient Greeks, the founders of Western civilisation. The ancient Greeks were not only adroit logicians and empiricists but also experimented in altered states of consciousness which probably triggered different ways of knowing the world. Hellenic rationalism never totally divested itself from supernaturalism as testified by the various ecstatic cults that are located in that period [265]. Even the Greek dynamism for intellectual inquiry did not obviate from its oracular tradition, so important to the Hellenic people. Indeed, the Greek notion of the psyche combined "natural and metaphysical elements" [266]. The anthropologist Gilbert Herdt observes that western conceptions of selfhood have been influenced by ancient Greek notions of the self [266]. Among the Homeric Greeks notions of the "Other" were associated with the psyche. This

"Other" could manifest as dream phenomena or a shadow image [267]. The Homeric Greeks did not ascribe any abstract duties to the psyche. Rather intense states were often ascribed to outside agencies which could affect emotion and behaviour [266]. For example, the frenetic ecstasies of the early *Bacchae* (followers of Dionysus, Hellenic god of wine) or the inspired madness of the oracle of Delphi were attributed to external divine forces. This belief was maintained through to the time of Socrates. Thus, in the *Phaedrus*, Socrates states that madness is divinely inspired whereas soberness is mundane [268]. The eighth century poet, Hesiod, described the "Other" in terms of a guardian of humans (*daimones*) who wandered throughout the earth in invisibility. Some of these *daimones* could act on their host's behalf to wreak punishment to a particular family, and were given the special title of *Alastor* [269].

Since the Mycenaean period the Greeks had various cults devoted to understanding altered states of awareness. The Eleusinian Mysteries which played an important part in classical Greece had been developed by the Mycenaeans by 1500 BCE [270]. Many deities of the Greek pantheon stemmed from Mycenaean deities. Prominent among these was Poseidon who had originally been a chthonic deity and associated with the underworld [271]. The Greek penchant with the chthonic powers had already been established by the Mycenaeans. Other deities which were connected with the underworld during this period were Demeter and Persephone who the Mycenaeans referred to as the 'two goddesses' [272]. The chthonic deity Dionysus the god of wine, madness and ecstasy was also worshipped by the Mycenaeans, although, Dionysian cults had also existed in Minoan Crete [273] and in Thrace [274]. He was probably originally a primordial nature deity; traces of Dionysus's shamanic origins are evident by his association with centaurs and satyrs. Second century mosaics portray Dionysus riding a leopard. In the Homeric hymns Dionysus has power over wild animals and transforms himself into a lion. The shapeshifting Dionysus also transforms a group of pirates into dolphins [275]. In an alternative tradition, Dionysus is the progeny of Zeus and the chthonic Persephone [276]. In Greek mythology, Persephone was the daughter of Demeter and had been abducted by Hades, god of the underworld. In her despair Demeter refused the earth to grow anything prompting Zeus to ask Hades to return Persephone to be reunited with her mother.

The Greek myths assisted in the development of various traditions during classical Greece. Two major traditions among the ancient Greeks became pivotal to their civilisation; these were the Delphic oracle and the Eleusinian Mysteries. The three traditions were connected with polyphasic modes of consciousness and assisted in the high creativity of the Greeks in science and the arts. The Pythia – the name given to the female oracle at Delphi was the most famous of the Greek oracles. Ancient Greece's history was at times intertwined with the Delphic oracle. The Pythia would always fall into altered states of consciousness, probably with the assistance of inhaling earthly gases which caused hallucinations. Modern researchers point out that the chthonic fumes which the Pythia inhaled probably contained high levels of ethylene which may have triggered trance and euphoric states [277]. Greeks believed that the Pythia mediated messages from the God Apollo. Her utterances were held in high esteem among a male dominated Greece. She was alleged to have foretold that only wooden palisades would save Greece from the Persians in 480 BC. Themistocles interpreted the prophecy to be ships and thereby sent the Athenian fleet to triumph over the Persians. The Pythia was one of several avenues which the Greeks had engaged in altered states of consciousness.

While the Delphic oracle had a high place in the Hellenic mythopoeitic imagination, the Eleusinian Mysteries provided a collective polyphasic experience. The Eleusinian Mysteries which were held every five years consisted of a series of collective rites and ceremonies. Greeks of all ages and professions came to be initiated in the Eleusinian Mysteries were connected to the Goddess Demeter who is associated with fertility and agriculture. So the Mysteries were probably associated with the themes of life, death and transformation – common themes found in shamanic rituals. Parts of the initiatory rituals comprised fasting and taking of a mystical brew called *kykeon*. Some scholars have argued that *kykeon* contained psychotropic substances. What *kykeon* was made of remains a mystery. Theories range from *kykeon* containing ergot, a barley fungus, opium, or Syrian Rue (*Peganum harmala*), a plant common in the Mediterranean region having anti-depressant functions [278]. An interesting connection between the Delphic oracle and the Eleusinian Mysteries is their link with chthonic powers which are important in shamanism.

For Jung, chthonic powers were associated with the unconscious – the place of psychic creativity. The universality of chthonic deities found throughout human cultures testifies to the power of the unconscious in directing the human imagination. For the Greeks the chthonic powers were a source of power and origination. Achilles was baptised in the underworld river Styx while Odysseus travelled to the underworld in order to return to his home. And then there is Heracles who had wrestled with Thanatos, son of Hades, and who had overpowered the three headed dog of the underworld, Cerberus. The Titans who had ruled the universe and had been defeated by their cousins, the Olympians, were imprisoned in Tartarus, the lowest level of the underworld. The Titans represent the primordial forces which cannot be controlled by human devices. They may also be associated with the atavistic aspect in humans - our innate instincts which need to be controlled by culture. However, it is the atavistic nature in humans which is a source of creativity and insight. In Greek mythology it is the Titan Prometheus who gives humans fire, while the Titaness Themis presides over divine order and social custom.

Ancient Greek shamanism offers a way for logic ridden Westerners. The Greeks acknowledged Apollonian (logic, reason) and Dionysian (intuition, dreams, trances) aspects of the human psyche, knowing that this pair worked in tandem in human nature to produce all kinds of creativity. The Greeks were caught in a tension between Dionysian mythical and magical tropes and "Apollonian rationalism" [279]. This friction underlined Greek civilisation and underlined the transition from Greek mythical to rational modes of thought during the sixth century B.C.E. when the first of the Greek naturalists such as Thales, Anaximander, Anaxagoras and Pythagoras started to think of the world as governed not by gods but by natural principles. The shift is marked in Greek mythology by the slaying of Python (representing the feminine and intuitive aspect found in divination) by Apollo (representing patriarchy and rationality) [279]. Ironically, the Delphic oracle was connected with Apollo among the classical Hellenes.

The Greeks understood that monophasic awareness was lacking without polyphasic experience. Greek mythology is replete with this polyphasic urge. In Oedipus Rex it is the prophet Tiresias who makes known to the ill-fated Oedipus his heinous sins. Greek myths also understood the relationship between women and their predilection towards intuitive modes of awareness. The Greeks venerated the female principle since Neolithic times. In Greek thought, the female principle was associated with fertility, power and prophecy. While the chief Olympians (Zeus, Poseidon, Hades) were males, pre-classical Greek deities venerated the goddess. In Greek mythology, the female principle of prophecy is associated with the Gracae – three ancient sisters who shared one eye and were chthonic deities. The names of these sisters were Enyo (the warlike), Deino (the terrible, and Pemphredo (the one who evokes alarm). In the Iliad, Enyo – one of the sisters led the Trojans into battle alongside Ares – the god of war. The Gracae probably embody the principle of vital power of the non-human world as they are also associated with animals. According to the playwright Aeschylus the Gracae had the heads and arms of old women and the bodies of swans. Another female trio were the fates who controlled the destiny of both humans and the gods. So great was their power that not even the gods could change destiny once it had been ordained by the fates. The fates are also the daughters of *Nyx* (night) who resides in a cave and gives oracles.

The birth of Athena is also relevant here since according to Greek myths she burst out from the forehead of Zeus – a virgin birth. Interestingly, this myth exemplifies the shapeshifting leitmotif of the Greeks, for Athena is a metaphorphosed Metis – a Titaness and first cousin of Zeus, who was turned into into a fly and was swallowed by Zeus. This fly then flew within Zeus's head until it was released by Hephaestus by cleaving the former's skull, and releasing the fully formed Athena. It is Athena who mentors Odysseus who encounters his own shapeshifting saga when his companions are turned into pigs by the immortal witch Circe. In these two myths it is females who play the role of shapeshifters – possibly an allusion to the creative powers of the feminine. Of course, this kind of mythopoietic thinking is evident in cradle civilisations such as Egypt which the Greeks had heavily borrowed from. Egypt's pantheon is studded by various theriomorphic gods such as the crocodile goddess, Ammut – who punished the wicked, the benign cow goddess Hathor who is also linked to augury, and the canid god of the dead, Anubis, who is likened to Cerberus - the three headed guardian dog of the Greek underworld. We may deduce that the sheer array of theriomorphs in ancient

Egyptian religion epitomizes its shamanistic heritage.

In Asian religions such as Hinduism, Jainism and Buddhism, the shapeshifting metaphor is typified by various snake deities widely known as *nagas*. Similar to the Muslim jinn, the *nagas* are associated with terrestrial, watery and chthonic regions. According to Hindu myth, the *nagas* were banished to the underworld by the creator god Brahma, where they have a protective role. The *nagas* are worshipped in southern India and are linked to human fertility. In Buddhism, it was the *naga* king Mucalinda who sheltered Buddha during his enlightenment meditation. Later on the *naga* shapeshifted as a human and paid Buddha homage. In Hinduism snake symbolism is linked to universal and generative power. For example, *kundalini* (primeval energy) is conceptualised by Hindus as a sleeping coiled snake at the base of the spine which can be awakened by meditational practices. Having been stirred, *kundalini* ascends along the spinal column (*sushumna*) arousing psycho-physical centres called *chakra* (seven in all). The Greek idea of the cosmic snake - the *ouroboros*, that eternally devours itself, was probably inspired by Vedic civilisation. Carl Jung explained the *ouroboros* as a principal mandala - a cosmographic depiction of the human psyche. As a regenerative principle, *kundalini* is quintessentially feminine, and it is this Vedic notion of the creative powers of the feminine which underpins Hindu religion.

The Greeks were probably intuitive about the prescient powers of females. However, in modern times the verdict is still out whether women are more precognitive than men. Research was conducted in the 1990's among a Scottish population to ascertain if women were more psychic than men. The survey uncovered that women had experienced more psychic episodes [280 - 282]. These findings emulated a nineteenth century study that stated that women had reported more than men that they had psi experiences [283]. More recently an internet survey had been conducted involving more than 15,000 people. Before the commencement of the experiment 78% of women classified themselves as being more intuitive while 58% of men made a similar claim [284]. The experiment was developed by the English psychologist Richard Wiseman, and consisted of photographs of smiling people. The aim of the experiment was to decide whether the smiles in the photographs were genuine or not. It turned out that men were more able to identify the genuine smiles than women by 1%. Furthermore, men detected 76% of fake female smiles, while women detected 67% of fake male smiles [284]. When it comes to paranormal phenomena men are just as interested as women are. The resurgence in belief in UFOs and Bigfoot among American males testifies to this. However, male interest in paranormal phenomena tends to be a means to discovery as evinced by the number of UFO and Bigfoot hunters [285]. This modern hunting behaviour is spurned by the neurotransmitter dopamine which plays a role in pleasure, reward and sexual desire. In men, the need for a dopamine rush has been implicated in high levels of extra-marital affairs. Alternately, women tend to believe in paranormal phenomena as a means of self-development [285]. To what degree biological and social mechanisms influence paranormal seeking behaviour in men and women is still uncertain [285].

We think that the Greeks were certainly onto something by their need to experience different kinds of altered states of consciousness. The Greeks had probably known that the repression of polyphasic modes of awareness was futile and led to the debilitation of the psyche. On this theme, the noted 20th century psychologist Carl Jung stated that any attempt by the individual or society to repress "reactions of the unconscious" were doomed to fail [286]. His ideas of the symbolic manifestations of the unconscious in the cultural realm have prompted an enormous amount of enquiry into previously forbidden domains such as dreams, possession, and ecstatic states.

Recent research indicates that analytical thinking that is prized in western civilisation may decrease intuitive religious based thinking. The thesis asserts that while intuitive thinking is based on mental shortcuts towards attaining fast results, analytical thinking is more protracted [287]. Could it be that shamanism followed intuitive thinking because of its ability to yield quick information which was needed in ancestral environments? We have seen that intuitive modes on knowledge provide unique ways for accessing knowledge about the self and the environment. This may be one reason why the otherwise logic ridden Greeks persisted in intuitive ways of knowing. In any case, it may be detrimental for modern humans to undermine this kind of knowledge. The shaman in all of us needs to be recognised as a significant human legacy for informing present and future humankind.

# CHAPTER 10

# What of the Future? Getting Serious About the Brain and Cognitive Enhancement Technologies

**Abstract:** From science fiction movies and books to the world of augmentation, modern humans are fixated by the future. What is in store for us? This chapter offers how future humans may be enhanced. Nowhere is this more evident than in our desire to augment the human brain. The exponential rise of biotechnology in the last thirty years has provided a necessary platform for future enhancement therapies to come into being. We explore various kinds of brain augmentation, present and future, while cautioning their unknown consequences for future humanity.

**Keywords:** Brain-machine-interfaces, Chimpanzee, Dopaminergic systems, *Homo erectus*, Kurzweil, Linden, Nootropic, Optobionics, Previc.

## A BRAVE NEW WORLD

In the last two decades there has been an exponential increase in cognitive science and neurosciences that has increased our understanding of the brain. New technologies such as brain imaging and molecular biology have provided a new picture of neural processes, especially in terms of how humans may experience the world. However, this is early days. Our pioneering ventures into the brain promise further insights. Whereas physics pathed the way for the nuclear age during the twentieth century, neuro-biology will probably lead humankind into the twenty-first century and beyond.

A cognitive enhancement may be defined as a biotechnological push towards augmenting human cognitive abilities beyond their biological constraints [288]. A therapeutic intervention on the other hand attempts to correct specific pathologies or defects [288].

**Maciej Henneberg & Arthur Saniotis**

The distinction between therapies and enhancements is becoming progressively unclear [289]. For instance amphetamines are more often recommended for enhancing mental functions. Furthermore healthy individuals are using SSRis (serotonin reuptake antidepressants to feel better [290].

Modern neuroscience research is finding out how neurones respond to sensory stimuli [291]. Future research will aim to discover how language, emotion and memory are facilitated by brain physiology. Technologies improving cognitive functions may extend outside of just therapeutic applications to actual brain function enhancements. Such enhancements may profoundly impact on future societies in ways which we cannot imagine. One possible scenario may be the use of neuro-prosthetic devices for enhancing intelligence and sensory perceptions, or specially designed pharmacological substances in combination with virtual reality technologies for enhancing mood and awareness.

Ray Kurzweil important work *The Age of Spiritual Machines* (2000) [224] highlights the need for improvement of brain functions. Although human brain is good at parallel processing (some 100 trillion synaptic connections between about 100 billion neurons) it is not a fast computing machine, it can only do 12,000 calculations per minute, as compared to billions of instructions per second that a modern computer processor can perform. The futurist Ray Kurzweil even goes as far as to postulate the DNA-based biological evolution may be abandoned. He believes that human-engineered brains using nanotechnologies will be many times superior with greater volumes of memory and faster recall. He notes that thorough research into our neural circuitry will eventuate in humans imitating and exponentially improving upon the brain's evolutionary design [224]. To engineer human brain it is necessary to understand its genetic, epigenetic and physiological characteristics produced by long evolution in a particular set of environmental circumstances. These characteristics underlie workings of the human mind. A re-design of our brain by engineering methods may significantly transform what we consider human mind. More importantly, there is a range of variation in human brain functioning. To ignore this variation is tenuous. Just what is perceived as the 'typical brain' may lead to terminal collapse of human brain functions because engineers using their imperfect, primate brains cannot predict everything that the human brain may be required to do in the future.

**What's So Good About the Human Brain?**

Contrary to much scientific speculation the complexity of the human brain is not exceptional in the world. This is mere human hubris. Sure, a lot has been discussed about the brain's exceeding neural complexity. This is undeniable. Composed of about 100,000 million neurons, with approximately 10,000 synaptic connections each, the human brain produces superlative cognition. WRONG! This kind of spiel is not only common but erroneous. What studies in evolutionary medicine show us is a brain that is inefficient, clumsily put together and vulnerable to a litany of neurological and psychiatric disorders which are mainly absent in our primate cousins. Here are a few examples of the brain's 'botchjobs'.

The human retina is positioned at the back of the eye, close to the optic nerve. However, the light sensitive, cylindrical shaped retinal cells lie at the bottom, a less than optimal design. Moreover, light photons have to traverse through various optical structures in order to get to the retinal light receptor cells – analogous to going through a maze before reaching the destination point [292, 293].

The Amygdalo-Hypothalamic circuit conjoins two major limbic structures. However, as befitting its central role it is poorly designed. Instead of following a linear axis, the circuit begins at the amygdala and follows a circular trajectory around other limbic structures until reaching the hypothalamus. This is probably an 'afterthought' by evolution due to limited brain space.

The brain's 'ad hoc' design is the central theme of neuroscientist David Linden's book *The Accidental Mind* (2007). His book deconstructs the myth of our superlative brain. For Linden, the brain is composed of several design deficits, a principal one being the inefficiency of neuronal communication. In short, neurons do not converse well between each other and have a tendency to leak. In fact, we are told that axonic release of neurotransmitters only occurs approximately one-third of the time. That is a big miss out rate. To overcome this problem, evolution simply increased the number of neurons in the neo-cortex [294].

However, as elegant as Linden's argument is, it overlooks neuro-hormonal regulation as a possible key driver of intelligence rather than focusing on brain anatomy. For example, the role of the neurotransmitter dopamine. In an interesting theory posited by psychologist Fred Previc, the dopaminergic system (DAergic) has played a central role in human brain evolution. According to Previc, during the Pleistocene period food procurement of ancestral humans, such as hunting and foraging, was physically demanding. This resulted in changes in thermo-regulation which in turn stimulated production of dopamine [14].

Secondly, during the Pleistocene period ancestral humans began to include seafood in their diet, thereby increasing the omega- 3 long chain polyunsaturated fat (PUFA) docosahexaenoic acid (DHA), which further expanded the DAergic. Consequently, the inclusion of seafood derived DHA may have increased levels of both dopamine and the thyroid hormone thyroxine (T4), necessary for regulation of human cognitive processes (*i.e.* language fluency, working memory). It is possible that DHA was positively selected in later *Homo* for its neurotrophic benefits, which may answer why modern humans are seemingly reliant on DHA for proper brain development and health.

Continuing from Linden, a major biological default of the brain is that humans learn *via* imitation even when imitative behaviours may have deleterious costs in the future. An experiment conducted by comparative psychologist Victoria Horner was based on 3-4 year old children and wild chimpanzees from an African sanctuary having to retrieve a piece of candy from a puzzle box. Horner demonstrated the 'required' actions for retrieving the reward. Both opaque and clear versions of the box were used. Both the chimpanzees and the human children successfully retrieved the candy from the opaque box by reproducing the action sequence of the demonstrator. When presented with the clear box "causal information was available"; it was evident that some of the sequential actions were unnecessary, however, all of the children still performed them. In contrast, the chimpanzees only performed the required actions for retrieving the reward while skipping the irrelevant actions. An uncomfortable conclusion from this study is that humans learn by imitating others, even when what is learnt can be maladaptive or deleterious. This predilection towards imitating others seems to be carried over to all stages of human life. Human attentiveness towards following the actions of others, especially 'authority figures' characterises a biological default in the brain/mind, and is an unfortunate and necessary trade-off of culturally based behaviour [295].

This imitative propensity of humans tends to supersede human novelty, the latter being extremely rare. Indeed, the brain has not evolved to be really good at novel thought. In ancestral environments it was not novel thought, but rather, tried and proven survival behaviours which increased the fitness of *Homo*. Such strategies would have invariably been passed down. Ability to memorise, was thus, crucial as survival depended on it. The survival value of mnemonic based intelligence, so valued by modern humans, probably arose during the time of *Homo erectus* (1.8 Ma) where there was an expansion in social and technological complexity. However, even the prodigious mnemonic memory expressed in prodigies and autistic savants is paralleled in other animals. There are many examples amongst avians which can memorise locations of thousands of seed caches, an incredible feat considering their 'smallish' brains. The most prodigious of these avian seed storers is Clark's Nutcracker (*Nucifraga columbiana*) which has been reputed to store over 30,000 pine nuts in over 2 thousand locations, covering 100 square mile area. Incredibly, this bird can memorise approximately 70% of its seed caches. Not bad, considering most of us cannot even remember what we had for dinner 4 days ago. How can this bird and others of the same ilk perform such mnemonic feats? Probably because of their larger hippocampus region, a structure which is involved in spatial memory. Scientists know that the hippocampus is implicated in working spatial memory in these avians, as lesions to this area drastically reduce their ability to recall locations and no capacity for re-learning them [296].

Even chimpanzees (*Pan troglodytes*) seem to have better memories than humans in remembering numbers, something which may be surprising since chimpanzees do not have a developed numerical culture. In 2004, a test was performed using chimpanzees and humans. The test was composed of numerical sequences (using the Arabic numerals from 1 to 9) appearing momentarily on a screen. The nine numerals were then transformed to white boxes. The participants then had to remember in which boxes the correct numerals were shown. Five out of the six chimpanzees outperformed their human participants in both accurately selecting the numerical sequences and in shorter duration time. It should be pointed out that the chance for correctly remembering the numerical sequence is 1/362,880. In contrast, the accuracy of the human participants' as a function of hold duration worsened during the trials, something which was not evident in five out of the six chimpanzees [297].

At this point, you may conclude that if the brain is so poorly designed then why have humans become the dominant species on the planet? In answering this we must examine what the brain has an ability for – symbolic language and technology. Indeed, it is usually these two capacities of the brain which humans tend to distinguish themselves from other animals. Again, this kind of human exceptionalism is self-deceptive. A plethora of ecological and anthropological studies reveal the complexity of non-human communicative systems and behaviours which were virtually unknown a few generations ago. Our knowledge of the animal kingdom has not only improved but we have a fairly good idea of the origin of human cognition and behaviour. For example, altruism and empathy, those most humanlike of qualities, are found in many animal species. Humans, the great apes, elephants and cetaceans may possess more evident forms of altruism due to their brains containing spindle neurons (otherwise known as Von Economo neurons) which play an important role in social awareness and cognition.

Additionally, there are many kinds of intelligence evident in nature such as social intelligence, kinaesthetic intelligence and sensory based forms of intelligence. Each kind of intelligence has adapted to the environment and conferred fitness value to individuals. It is ludicrous to compare or judge which kind of intelligence is better. In humans symbolic and technological based intelligence are privileged because they enhanced and expedited cultural evolution in ancestral *Homo*, to the point that humans ever since have been reliant on them. Social anthropologists note that humans who have not been acculturated lack the ability to think in symbols, and sometimes cannot even walk in a bipedal fashion. Even walking on two legs it seems is a learned, and not, an innate behaviour. Organisational features of animal brains may provide insightful hints into their types of intelligences. For example, the cortical arrangement of bottle-nosed dolphins reveals that the visual and auditory brain centres are adjacent to each other. This arrangement is in contrast to primate brains where the visual and auditory centres are located in occipital and temporal lobes. It has been suggested that this cortical adjacency in dolphins may provide a more integrated cross-modal, sensory processing ability which may inform complex social patterns in these cetaceans

[298].

While the human brain has been able to develop complex social patterns, technology and science, its ad hoc structure needs to be recognised by the public. If the human brain, is therefore, limited due evolutionary constraints, it may be reasonable to invent methods to augment brain function. This will be theme of the next sections.

## BRAIN-MACHINE INTERFACES

Eugene Thacker maintains that we are now remediating, enhancing the human body and we will start soon enhancing our brains by inserting chips into them. Brain remediation is a form of embodiment. There is an ongoing scientific effort to map the brain using biomedical technologies. Thacker thinks that the future scientific effort will be directed towards re-building the body, including the brain. In this way they will be redefined. Medical sciences will see the brain as both a biological organ and an object for technological manipulation [299]. The brain of the future will be a focus of biomedia, that is a complex of technologically advanced methods based on scientific knowledge [299]. Scientific activity of the future will tend to optimise human biology by technological interventions. However, without respecting biological variation human biology cannot be optimised. Humans have learned this *via* hard lessons such as infectious diseases. 'Optimisation' means unification of function. However, this could be evolutionarily precarious since it requires a source of variants for adaptations to unpredictable, external conditions. Bearing this in mind, before we get carried away, any brain simulation will have to understand the prodigious neuro-hormonal processes and neural synaptic networks; a task which will need far more than technological sophistication and neuroscientific understanding. Even the noted Stanford University neurologist, Robert Sapolsky, admits that our present understanding of how the brain works is negligible. For this reason, many scholars have concluded that we may never be able to fathom the brain's complexity. While any redesign of the brain is not solvable in the near future, cognitive enhancements in the way of brain-machine interfaces, cosmetic neurology and nanotechnology may offer some promise, but at a cost.

Development of technologies, including information technology, neuroscience and nanotechnology enable construction of brain-machine interfaces that may increase sensory and cognitive abilities of future generations. Artificial devices have already been integrated into primate brains [300]. Amplifications of sensory input and nerve signaling within the brain are currently investigated [290]. There is a need to improve our understanding of coding of sensory information and memories. This may assist people who suffer from sensory deficits and those who are paralysed [290]. Also, there are possible military applications of brain enhancement technologies where greater sensory acuity and fast decision-making are essential [301]. Despite progress of brain-machine interfaces research in recent decades, the devices of this kind are still at a nascent stage and their actual applications in humans are expected in the future [302].

Enhancement of our neurological performance by the use of artificial devices will be increasing in forthcoming years. Approximately 59 thousand individuals have received neurological prosthetics after 2002. Nineteen new devices were accepted for use in the decade beginning 1994 [303]. As early as 1979 an auditory implant was inserted into the brainstem [303].

Brain-machine interfaces translate activity in neurons into computer-understandable signals that may be used in specific prostheses [304]. Brain-machine interfaces use various signals coming from brain motor cortical areas, triggering neurological activity that stimulates upper limb movements [304]. This necessitates that a brain-machine interface learns to translate neurological patterns into signals that trigger movements [305]. The success of brain-machine interface-machine interface neural command sets is based on their capacity to retrieve information of neural commands in a time sequence [304]. Codifying nerve signal sequences for brain-machine interface practical uses is still being a subject of ongoing research. Although complicated to produce, a brain-machine interface programming once constructed can reflect information contained in nerve cells so that it can be mapped [304]. Brain-machine interfaces and brain-computer interfaces (BCIs) can restore functions in people with sensory or motor deficits due to injuries or diseases [306]. This approach is based on the principle of electrical activity of neurons being converted by electro-physical methods to signals readable by electronic devices [300].

In the next decade decisive steps in the brain-machine interface construction will be made through intra-cranial invasive implements and *via* non-invasive placement of electrodes on the skull surface. In clinical conditions it is possible now to produce situations that unify prosthetics with the body of a patient [307]. The time will come when it will be possible to influence the neural processes inside the brain by electronic devices operated by computers. This could improve intelligence and allow control over emotions [291]. The future effectiveness of brain-machine interfaces results from the fact that they act in a digital fashion that produces better precision than chemical analogue operation of pharmaceuticals that alter brain functions. Brain-machine interfaces can be placed in well-defined locations in the nerve tissue, avoiding side-effects and acting quicker than drugs [291].

According to Kurtzweil brain implants may be able to transmit digitally created artificial environments producing new sensations and new stimuli that are different from anything experienced naturally by the body [224]. These novel sensory inputs may significantly improve the intelligence and the knowledge of the world [224]. Moreover, spiritual experiences modulated by neural circuits may be increased if cortical areas such as frontal lobes are externally stimulated [224]. This forecast by Kurzweil is in agreement with Newberg and Lee's statement that neural implants can be inserted into those areas of the brain that control religious experiences and enable construction of myths [308]. Use of different kinds of brain-machine interfaces in the future will inevitably lead to experience of virtual realities of new kinds. According to Greenfield robots and other advanced digital machines will interact with humans in virtual environments [232].

According to Kurzweil brain-machine interfaces will become so widespread in the future that humans will not be able to function without them in the world filled with complex technologies. This author describes how computational devices will be comfortably integrated with human brains to the extent that their presence will be perceived as a normal part of human life. Digital brain supplementation will progress in steps until the structure of human intellectual processes will depend on the interaction with artificial devices. Kurzweil predicts by 2029 we will see optimization of high bandwidth brain routes [224]. This may enable the circumvention of some brain areas and improvement of these regions by prosthetics [224]. In his opinion brain-machine devices will become so universal that they will inform future human microevolution:

If we can provide the brain with speedy access to unlimited memory, unlimited calculation ability, and instant wireless communication ability we will produce a human with unsurpassable intelligence. We fully expect to demonstrate this kind of link between brain and machine [309].

Andy Clark similar to Kurzweil thinks that novel coupling of prosthetic devices with the human brain may improve health [310]. For instance, levels of neurotransmitters can be sensed by neural implants and these readings used to regulate the secretion of neurotransmitters in synapses. This may optimize the neuro-hormonal regulation thus avoiding depression and psychotic illnesses. According to Clark, brain-machine interfaces may connect global positioning systems (GPS) directly to our bodies and help us to interact with our environments. The merger of prosthetic devices with GPS may have various cartographical and aviation applications. Clark also proffers that prosthetic implants may safely be connected to neural tissue due to neuroplasticity [310]. Whether neural prosthetics will affect human endocrinology is still unknown territory. This brings forth the question as to what degree may neural devices inform human physiology.

Greenfield suggests, similar to Kurzweil, that people in the future may have kaleidoscopic levels of mentation; a thought-provoking idea [232]. Work on brain-machine interfaces has been pioneered by Edward Schmidt (1980). He proposed that electric activity of the brain can be used directly to deliver voluntary motor commands to muscles in cases of spinal cord lesions [311]. Leading neuroscientists using BMI, Phillip Kennedy and Miguel Nicoleis developed this innovative concept of Schmidt. Kennedy also suggested that the link between brain and periphery of the body may be reversed so that brain implants will be able to deliver new information to the brain, thus augmenting intelligence. In the future, great apes may be recipients of neural prosthetics which could give them the ability to speak [291]. A neurotrophic electrode was patented in 1989 by Kennedy. He tested this device on primates. The neurotrophic electrode had the ability to convert neural signalling to electromagnetic waves and to increase them by some 10,000 times. The radio signals generated allowed transmission of signals to a computer. A brain stroke victim, Johnny Ray was the first recipient of this augmenting device that was inserted into the brain area controlling the left hand. He was instructed to think about using his left hand to operate a computer mouse. This thought of Ray's produced cortical activity that controlled movements. After half a year of training Ray could move cursor on the computer screen by just thinking about it [309]. A monkey named Belle had in 2000 one hundred wires placed in her motor cortex by Nicolelis. These were connected to a computer. She operated a joystick. Signals from her brain were transmitted *via* a computer to a robotic arm which copied Belle's joystick movements.

Expanding on this experiment, 700 electrodes were inserted into rhesus monkey's motor cortex areas. Monkeys were trained to use joysticks. Electrodes fed their signals to a computer that translated the brain firing patterns into movements of a robotic arm [291]. Signals from head surface sensed by 64 EEG electrodes were used by human subjects to control a computer cursor in a development described by Wolpaw and McFarland in 2006 [312, 303].

In an experiment conducted in Kyoto, Japan in 2008, electrodes were implanted into the brain of a rhesus monkey called Idoya. A video camera was used to monitor her movements. The experiment showed that a computer was able to predict Idoya's leg movements approximately one second before she actually had performed them. A remarkable feat of high tech prescience with various future applications (*i.e.* forecasting brain activity of an intended target) [309, 313].

Technical problems with the use of brain-machine interfaces still exist. An immediate problem is the need for surgical intervention to connect the machine to the brain [303]. Besides trauma it also produces risk of an infection that may lead to meningitis. Another problem are qualities of an EEG signal. These signals are transferred slowly, at about 20-30 bits/min. Precise location of electrodes may be influencing their collection of signals from the cortex. Such location still needs optimisation. Moreover, the limited number of electrodes that current technologies can implant means that only a limited number of neurons can be tapped so that a full set of signals controlling movement may not be adequately captured [303].

Since the beginning of the 21$^{st}$ century more than 60 thousand patients had neurological devices implanted [303]. Amongst those common were cochlear implants converting sound waves to electric potentials in the electrodes inserted into auditory nerves. Auditory brainstem implants, which are a newer solution to the problem of deafness, have been in use since 1979. This implant helps patients who lost function of auditory nerves [303]. Visual prosthetic devices are less advanced, though a number of them are being developed, such as Optobionics or the artificial retina [303, 314].

The artificial hippocampus received a lot of public attention as early as 2003. Theodore Berger developed this device at the University of Southern California (UCLA). His team excited a hippocampus of a rat with electric stimuli to map outputs from nerve cells [308]. This enabled the construction of a mathematical model of changes in various cell layers of the hippocampus. This model was transferred to a digital chip. All this was done in order to eventually create a practical application that would consist of a chip located externally on the skull that will have two sets of electrodes inserted on both sides of the deficient hippocampal area [308]. This arrangement could be applied in cases of hippocampus being damaged due to epilepsy, psychopathology, stroke or Alzheimer's dementia.

## NON-MEDICAL USES OF BRAIN-MACHINE INTERFACES

Chief among future other than therapeutic uses of brain-machine interfaces will be for military purposes. For the past decade the American military has been pursuing the effective use of non-invasive or semi-invasive brain-machine interfaces. While results of various brain-machine interface projects await, it is tantalising to predict future military applications. Here, Hollywood offers an imaginative input. The 1982 movie Fire Fox was based on a fictional super jet of the same name which Clint Eastwood (playing an American secret agent) must steal before the Russians can use it in a cold war arena. While the Fire Fox is superbly crafted with space age instrumentation and speed, its deadliest weapon is thought control. The Fire Fox can establish a mind link with the pilot with devastating results. If you think that this is science fiction then think again at what brain-machine interfaces have been able to achieve so far. As clinical studies have shown invasive brain-machine interfaces can manipulate matter with mind. Of course, eastern mysticism has for thousands of years proclaimed how consciousness informs all universal matter. However, this idea is new to western science.

Psychological problems such as post-traumatic stress syndrome may in the future be treated with brain-machine interfaces [315]. During the last decade the Defence Advanced Research Projects Agency of the USA (DARPA) has been working on various projects using non-invasive and semi-invasive brain-machine interfaces in order to enhance soldiers' combative capabilities. One project which resembles the Fire Fox movie is called "Silent Talk". The basic idea is that soldiers may communicate with each other on the battlefield through EEG signals of intentional language [315]. In this way, verbal speech is unnecessary. This kind of silent talk may prove beneficial for reconnaissance missions or combative manoeuvres requiring stealth. Silent Talk is a continuation of early warrior societies who used ingenious methods to communicate over long distances *via* smoke signals, drumming, mirrors, or the humble messenger pigeon. A second DARPA project is based on brain-machine interface binoculars that respond subconsciously to a detected threat or target. It is called "Cognitive Technology Threat Warning System". The detection range of brain-machine interface binoculars is approximately 10 kilometres making them a portable and practical piece of military engineering [315].

Future generations of soldiers will probably be fitted with an array of brain-machine interfaces that are linked to external prosthetic devices. Brain-machine interfaces may work conterminously with special military body suites that will be able to gauge a soldier's vital functions and psychological states and relay these to medical centres. In this way, multiple soldiers will be systematically monitored within a war arena. This may be the next step in the non-military use of brain-machine interfaces in order to locate and control people in the future. Of course such use of brain-machine interfaces has considerable neuro-ethical implications such as infringement of personal identity and informed consent [316]. For example, what if brain-machine interfaces may enable to influence human personality or function to emphasise certain valued behavioural traits? Will this surmount to a form of personality control by governments or elite agencies? Indeed, such use of brain-machine interfaces will force future humans to re-value personal freedom, something which individualistic orientated western societies may find difficult to do. In the wrong hands or regimes brain-machine interfaces could become an ultimate weapon of control. Fundamentalist movements could use brain-machine interfaces to enhance indoctrination and to make individuals into very efficient terrorists. For this reason, future non-therapeutic uses of brain-machine interfaces will need special monitoring. According to Maguire and McGee a third phase in brain-machine integration will use artificial devices for optimising transfer of information [317]. Although this phase is still in its initial stages it illustrates future potential of brain-machine interface applications as technologies become more sophisticated.

## WILL WE BECOME CYBORGS?

The advent of brain-machine interfaces and their future use may force humans to re-define where our humanity lies. The reason for this is that increasing use of brain-machine interface will blur the boundary between human and machine. Some may see that this boundary has been blurred for a long time. Millions of people are wearing various devices to enhance our lives; eye glasses, hearing aids, pace makers, bionic ears, metal plates and pins, heart valves, and artificial limbs. We are becoming increasingly dependent on prosthetic devices and will so in the future. It is easy to think of the body as machine as parts of it can be replaced. Descartes in the seventeenth century was influential in formulating the body as machine, an idea that has informed modern bio-medicine. Of course, the body is not a machine but an evolutionary organism of enormous complexity. However, the machine metaphor remains attractive since the body is being continually invaded by an ever increasing range of prosthetic devices. A question arises whether such forays into the human body will change our humanity? For many thinkers the answer is yes [317].

In terms of evolutionary processes we are beginning to see at present hybridisation of digital devices and people [318]. This phenomenon, according to Donna Haraway, can be observed in numerous aspects of peoples' interactions with others. Introduction of internet allows to replace face-to-face interactions with disembodied forms of socialisation physically replacing a human interlocutor with a machine. Computers now provide platforms for novel forms of personal experience such as cybersex, virtual relationships with non-existent personalities and conversations with cyborgs. Donna Haraway introduced a term "cyborgisation of *Homo sapiens*" to describe technologisation of human individuals. According to this author a blurring of body/machine boundaries is occurring. It may fuse different forms of human consciousness transmitted by different networks that will include both biological neural circuits and the artificially constructed electronic ones. Humans are already cyborgs Maheu tells us since many of us are immersed in the cyber world [319]. According to Haraway flexibility and worldwide reach of the digital world enables humans to transcend bonds of kinship, thereby freeing themselves from responsibilities societies impose in systems of kinship.

## COSMETIC NEUROLOGY

Cosmetic neurology is a practice of using pharmaceutics in healthy people to improve their cognitive functions. We can start this section from an example that illustrates major problems related to cosmetic neurology. Dr. Chatterjee [320] attends to a patient who is a businessman, an adult man planning to travel to Saudi Arabia to conduct business negotiations. The businessman wants to improve his competitiveness by learning Arabic. He asks the doctor to prescribe pharmaceuticals that will make his learning easier by improving cognitive abilities. Dr Chatterjee obliges by prescribing the hypnotic Zolpidem, a stimulant, Modafinil and a small quantity of Dextro-Amphetamine. The decision to prescribe this noradrenergic agonist is justified by the knowledge that Dextro-Amphetamine fosters neuroplasticity and improves memory thus supporting learning of languages [321]. In many societies there is a trend towards the use of pharmacological products to improve 'normal' brain functions [290]. Regulation of human behaviours by doses of pharmacological substances has a potential to become widespread.

Nootropic drugs are now popularly used in developed countries. In 2008 about 20% of survey respondents used such stimulants like methylphenidate, modafinil, and beta blockers [322]. Conducted in 2007 in United States survey indicated that some 1.6 million people in that country were using prescription stimulants [322]. Nootropic substances have been known for decades. However, only recently their use became widespread for cognitive enhancement among healthy people of various ages and professions About 18% of American university students often use modafinil [323] since it improves mental focus and alertness [288, 324 - 326]. Another drug, adrafinil is prescribed for relief of sleepiness, mood improvement and increase of mental focus, while methylphenidate (Ritalin) is increasingly used for ADHD. Both modafinil and adrafinil stimulate noradrenergic post synaptic receptors increasing glutamatergic transmission and activating orexigernic neurons which induce wakefulness, alertness and memory improvement [327]. Young professionals and university students increasingly consider modafinil as a 'lifestyle drug'.

The term "nootropic" has been coined in the 1970s by Guirgea who believed that pharmacology can engineer further evolution of the human brain [328, 329]. Dextro-amphetamine and methylphenidate stimulants are prescribed shift workers and students for improvement of alertness. ADHD in children is treated with Ritalin. More than 8 million children have been prescribed the drug. Modafinil is used to enhance spatial memory in adults [330, 331].

The widespread use of amphetamines as cognitive enhancers, in particular, is a case in point. Amphetamines such as methamphetamine, crystal methamphetamine, and ecstasy are synthetic stimulants [332], Amphetamine is an old drug first time synthesised in 1887 [333]. Its chemical relative methylphenidate was produced in early 1900s [334]. In 1930s, hyperactive children were treated with amphetamine [335]. It was also used for the treatment of narcolepsy [334]. During the World War II times allied soldiers improved their alertness taking benzedrine, while Japanese and German military used methamphetamine for similar purposes [322]. After the war there was a shift towards recreational use of amphetamines that were also used by dieters [334]. Methylphenidate was considered a milder stimulant since its introduction in 1956 [322]. Billions of amphetamine pills were prescribed in 1960s [322]. United States National Survey on Drug Use and Health conducted in 2007 reported that among Americans 12+ years old 5.3% had used methamphetamine at some point in their lives [336].

The 'club drug' ecstasy has been commonly used recreationally mainly by middle class suburban youth in developed countries [337]. Ecstasy is a synthetic amphetamine which contains the chemical compound 4-methylenedioxyme-thamphetamine (MDMA). Ecstasy may also contain other nasty compounds which have been added which may cause serious side effects [338]. In the popular youth culture, during the 'rave parties' alcohol, ecstasy and other drugs are commonly used. Ecstasy produces a number of after effects: numbness, nausea, muscle aches, dizziness, blurred vision, vomiting, stomach pains, disorientation, anxiety and panic attacks and irregular menstrual cycles [337]. Long term use of ecstasy leads to permanent depression and anxiety due to its destruction of serotonin receptors. Serotonin is an important neurotransmitter which is central to mood elevation.

Various kinds of drug treatments, such as dopamine agonists, are currently used to improve motor performance. They are known to improve motor learning by enhancing neural plasticity. Dextroamphetamine is used in therapy of stroke patients and brain trauma victims [339 - 343]. In the future new methods will be available to manage diseases such as Parkinson's and Alzheimer's. Progress in neuroscience, nanotechnology and molecular biology will contribute to development of cosmetic neurology. Attention-modulating drugs, *e.g.* Cholines-terase Inhibitors may be recommended to improve cognitive performance [344], modafinil prescribed for increased vigilance [326] and Atomoxetic for augmenting arousal levels [323] Neuropeptides will be applied to modulate emotional states [345] including Vasopressin and Corticotropin Release Factor (CRF) [346]. The 'cuddle hormone' oxytocin, fostering feelings of trust is presently being distributed *via* inhalators [347].

Mood modulating substances, including anti-depressants, are another major group of drugs used in cosmetic neurology. It is believed that some 20% of Americans have some form of depression [348]. Moreover, close to 50% of Americans experienced illnesses related to substance abuse or emotional problems [349]. In the last two decades treatable depression cases have increased five-fold [289]. Serotonin reuptake inhibitors (SSRIs) are currently major anti-depressants. Post-traumatic stress disorder and anxiety are treated with beta-blockers. Widespread use of antidepressants may create a danger of their use in reinforcing social conformity in the future. Popularity of such brain-chemistry-altering drugs to be used for manipulation of behaviour of healthy individuals should be seriously considered [350]. Addiction can result from common use of certain SSRIs (*e.g.* Prozac and Zoloft) [289]. The important new phenomenon of including neuroactive drugs into personal health management raises a question of how boundaries between healing and cosmetic enhancement should be defined [351].

## NANOTECH BRAINS

Many futures thinkers consider nanotechnology as a kind of holy grail. At present most nanotechnological designs are at the stage of theoretical considerations. Their applications are expected in the near future. There is a great deal of hype regarding how nanotechnology will transform future societies. Even if half of nanotechnological inventions come to fruition it will be a very different world from the one we know. Among advantages of nanotechnologies is their capacity to change objects at a molecular level. Nanotechnology may even restructure cells such as neurons in the cortex to produce neural pathways influenced by technology. Ray Kurzweil is interested in constructing nanotech brains because he is concerned about the present state of human brain development. In his opinion biological evolution is much too unreliable and slow to advance cognitive development. For Kurzweil, although the biological brain is a wonderful parallel information processing device, it is much too slow in information processing. The fact that the human brain is for Kurzweil a biological organ makes it far too restrained and clumsy for his liking in comparison to a super optimal computer which would be exponentially faster and mnemonically superior. Kurzweil thinks that human brain must be fundamentally rebuild, molecule by molecule. The re-built brain will not be subject to biological rules. Additionally, nanotech brains will have improved intellectual abilities and superior sensory capacities. The way those improvements occur is a matter of speculation. A technological brain will probably have a different type of integration of information that will result in a different type of consciousness. Such brain will be very different from what biological evolution has produced. It is difficult to say whether such a transformation of biological brain into a technological one will be beneficial. Our biological brain is a product of thousands generations undergoing evolution. Human brain-mind has been formed by complex ongoing cultural and biological processes. The nervous system with its autonomic parts (sympathetic and parasympathetic) is widespread across the entire human body and stimulated by environmental and symbolic cultural phenomena. For example, pleasure centres of the brain are shared by humans and other mammals which may explain why we experience empathy with other animals. Individual exploring multiple realities experiences increased neural empathy [183]. It is important to note here that the structure of the brain is reflecting a polarity between its own integrity and adaptation to the external world [183]. The structure of the brain is genetically determined by faculties which were called "neurognosis" by Laughlin [183]. Neurons competing with each other in "topographical competition" constitute neurognostic structures [352]. This significantly influences neural wiring of the brain because social experience constantly reforms neural pathways. A non-biological nanotech brain may not be able to mould similar neurognostic structure of neurons competing under varied experiences. The fact that brain is capable to experience simultaneously multiple phase states of non-ordinary consciousness such as trance, hallucinations, dreams and visions is a significant characteristics of the neurognosis. Would it be possible for nanotech brain to have similar multiple states of consciousness? Some futurists promise that we will be able to have a utopian life free from anxieties, psychoses, fear and mental illnesses when we use nanotech brains. Synthetic nano-designed neurons inserted in the cortex are expected to exponentially increase the parallel processing of information in the brain while at the same time providing links to the internet [353].

The popular futures thinker Nick Bostrom has envisioned the ability to download digitally the mind by replicating activities of the brain at a molecular level and then scanning the brain in order to create a neural simulation of it [228]. He states that a set of signals downloaded from the mind onto a digital chip or into a computer "mind file" will imitate an organic person because it will have the memories and the personality of that person [353]. Once a mind is electronically downloaded its backup copies can be stored for long periods, thus producing immortality beyond the usual period of the existence of the human body [229]. While such mind simulations may appear to provide immortality, a kind of enslavement may result from being trapped in the computer memory; an unmodifiable existence inferior to variable biological life. Any nostalgia for carbon based brains would be temporary as the mind simulations on file would be viewed as a superior option. Deriving mind simulations from activities of biological brains will be challenging since it calls for emulating the almost impossible to predict complex neural matrix. As advances are made in construction of virtual reality, computer mind simulations will improve their efficiency [354]. Nanotechnology will be the most important in achieving these improvements. For example, a planetary computer can perform some $10^{42}$ operations per second [354]. The entire knowledge accumulated throughout human history could be stored in this kind of a computer while using significantly less processing power [354]. A future civilisation is likely to have a large number of super computers not only storing limitless information, but also opening up new neuroscientific processes and improving cognition. The essay "The Homo Cyber Sapiens, the Robot Homonidus intelligens," describes future people as having brains enhanced by computer interfaces [355]. This will enable people to improve their understanding of the evolution of consciousness and explore origins of human intelligence [356]. An alternate conception of the future makes people to live in a virtual reality. According to the physicist Freeman Dyson, people of the future will use interface of brains and electronics to enable neurotelepathy [232]. Cyber network will transform input from the outside world into sensory perceptions. Immersion with the cyber world would be posited on a kind of neurotelepathy rather than brain intrusive technologies [232]. In the current century molecular engineering in neuroscience is expected to produce new neurotechnologies that will probe various aspects of brain functions transforming its operation in the future life. The neural environment will be slowly uncovered [232]. The future world will be transformed into a mental matrix. This trend is becoming visible already in our dependence on digital devices and computers. Machines may even empower humans to understand better their consciousness. In the future the boundary between fiction and reality will be gradually blurred ending up in a formation of a virtual world. The present network of traditional family ties will be replaced by internet-mediated relationships with persons who may not even exist outside of the virtual reality. Fictional persons with artificially produced life-histories may be able to create friendships with real people [232].

Artificially created relationships may become a popular phenomenon in the expanding virtual world. This future world of complex cybernetic machines is likely to profoundly alter our ideas of personal autonomy. People may become indifferent to their actual circumstances since their sensory desires would be satisfied by artificial technologies while the participation in virtual world will be preferable to real life. Human grasp on reality may be challenged in the world saturated with computers to such an extent that people will become docile actors in a dehumanised digital network [232]. It may take a hardship and lack of sophisticated technologies to re-ignite human initiative in search for alternative solutions to life and to understanding of the world. Hardship and poverty, however, should not be advocated as the means to change.

# References

[1]     Lovejoy CO. Reexamining human origins in light of Ardipithecus ramidus. Science 2009; 326(5949): e1-8.
        [http://dx.doi.org/10.1126/science.1175834] [PMID: 19810200]

[2]     Galik K, Senut B, Pickford M, *et al*. External and internal Morphology of the BAR 1002'00 Orrorin tugenensis femur. Science 2004; 305(5689): 1450-3.
        [http://dx.doi.org/10.1126/science.1098807] [PMID: 15353798]

[3]     Niemitz C. The evolution of the upright posture and gait - a review and a new synthesis. Naturwissenschaften 2010; 97(3): 241-63.
        [http://dx.doi.org/10.1007/s00114-009-0637-3] [PMID: 20127307]

[4]     Vaneechoutte M, Algis M, Kuliukas A, Verhaegen M, Eds. Was man more aquatic in the past? Fifty years after Alister Hardy: waterside hypotheses of human evolution. USA: Bentham eBooks 2012.

[5]     Schoeninger MJ. Palaeoanthropology: the ancestral dinner table. Nature 2012; 487(7405): 42-3.
        [http://dx.doi.org/10.1038/487042a] [PMID: 22763547]

[6]     Wall-Scheffler CM. Energetics, locomotion, and female reproduction: implications for human evolution. Annu Rev Anthropol 2012; 41: 71-85.
        [http://dx.doi.org/10.1146/annurev-anthro-092611-145739]

[7]     Gillman MW, Gluckman PD, Rosenfeld RG, Eds. Nestlé Nutr Inst Workshop Ser Recent advances in growth research: nutritional,molecular and endocrine perspectives 2013; 71: 89-102.
        [http://dx.doi.org/10.1159/isbn.978-3-318-02270-4]

[8]     Dominguez-Rodrigo M, Ed. Stone tools and fossil bones, debates in the archaeology of human origins. Cambridge: Cambridge University Press 2012.
        [http://dx.doi.org/10.1017/CBO9781139149327]

[9]     de Boer B. Investigating the acoustic effect of the descended larynx with articulatory models. J Phonetics 2010; 38: 679-86.
        [http://dx.doi.org/10.1016/j.wocn.2010.10.003]

[10]    Arensburg B, Tillier MA, Vandermeersch B, Duday H, Schepartz LA, Rak Y. A middle Palaeolithic human hyoid bone. Nature 1989; 338: 758-60.; Lieberman P. On the origins of language. New York, NY: Macmillan 1975.

[11]    Stephan CN, Henneberg M. Medicine may be reducing the human capacity to survive. Med Hypotheses 2001; 57(5): 633-7.
        [http://dx.doi.org/10.1054/mehy.2001.1431] [PMID: 11735325]

[12]    Saniotis A, Henneberg M. Medicine could be constructing human bodies in the future. Med Hypotheses 2011; 77(4): 560-4.
        [http://dx.doi.org/10.1016/j.mehy.2011.06.031] [PMID: 21783327]

[13]    Prufer K, Munch K, Hellmann I, *et al*. The bonobo genome compared with the chimpanzee and human genomes. Nature 2012/06/13/ advance online publication 2012; 1476-4687.

[14]    Previc F. The dopaminergic mind in human evolution and history. Cambridge, UK: Cambridge

University Press 2009.
[http://dx.doi.org/10.1017/CBO9780511581366]

[15]   Henneberg M, Saniotis A. Evolutionary origins of human brain and spirituality. Anthropol Anz 2009;
       67(4): 427-38.
       [http://dx.doi.org/10.1127/0003-5548/2009/0032] [PMID: 20440961]

[16]   Saniotis A, Henneberg M. An evolutionary approach towards exploring altered states of
       consciousness: mind-body techniques and non-local mind. World Futures: J Gen Evol 2011; 67: 182-
       200.

[17]   Saniotis A, Henneberg M. Future evolution of the human brain. J Future Std 2011; 16(1): 1-1.

[18]   Henneberg M, Thackeray JF. A single-lineage hypothesis of hominid evolution. Evol Theory 1995;
       13: 31-8.

[19]   Lee BS, Wolpoff MH. The pattern of evolution in Pleistocene human brain size. Paleobiology 2003;
       29(2): 186-96.
       [http://dx.doi.org/10.1666/0094-8373(2003)029<0186:TPOEIP>2.0.CO;2]

[20]   Eckhardt R. Human paleobiology. Cambridge: Cambridge University Press 2000; pp. 166-207.
       [http://dx.doi.org/10.1017/CBO9780511542367]

[21]   Henneberg M, de Miguel C. Hominins are a single lineage: brain and body size variability does not
       reflect postulated taxonomic diversity of hominins. Homo 2004; 55(1-2): 21-37.
       [http://dx.doi.org/10.1016/j.jchb.2004.03.001] [PMID: 15553266]

[22]   Brace CL. The stages of human evolution. Englewood Cliffs, new Jersey: Prentice-Hall 1967.

[23]   Beauvilain A, Watté JP. Toumaï (Sahelanthropus tchadensis) a-t-il été inhumé? Bulletin de la Société
       Géologique de Normandie et des Amis du Muséum du Havre 2009; 96(1): 19-26.

[24]   Franzen JL. Asian australopithecines? Hominid evolution: past, present, and future. New York: Wiley-
       Liss 1985; pp. 255-63.

[25]   Berger LR, Hawks J, de Ruiter DJ, *et al.* Homo naledi, a new species of the genus Homo from the
       Dinaledi Chamber, South Africa. eLife 2015; 4: e09560.
       [http://dx.doi.org/10.7554/eLife.09560] [PMID: 26354291]

[26]   Toro-Moyano I, Martínez-Navarro B, Agustí J, *et al.* The oldest human fossil in Europe, from Orce
       (Spain). J Hum Evol 2013; 65(1): 1-9.
       [http://dx.doi.org/10.1016/j.jhevol.2013.01.012] [PMID: 23481345]

[27]   Bermúdez de Castro JM, Arsuaga JL, Carbonell E, Rosas A, MartA-nez I, Mosquera M. A hominid
       from the lower Pleistocene of Atapuerca, Spain: possible ancestor to Neandertals and modern humans.
       Science 1997; 276(5317): 1392-5.
       [http://dx.doi.org/10.1126/science.276.5317.1392] [PMID: 9162001]

[28]   Arsuaga JL, MartA-nez I, Gracia A, Carretero JM, Lorenzo C. Sima de los Huesos (Sierra de
       Atapuerca, Spain). The site. J Hum Evol 1997; 33(2-3): 109-27.
       [http://dx.doi.org/10.1006/jhev.1997.0132] [PMID: 9300338]

[29]   Carbonell E, Bermúdez de Castro JM, Rosell J, Arsuaga JL. The first hominin of Europe. Nature 2008;
       452(7186): 465-9.

[http://dx.doi.org/10.1038/nature06815] [PMID: 18368116]

[30]   Delson E, Harvati K. Palaeoanthropology: return of the last Neanderthal. Nature 2006; 443(7113): 762-3.
[http://dx.doi.org/10.1038/nature05207] [PMID: 16971950]

[31]   Saniotis A, Henneberg M. Othering of neandertals. Before Farming 2010; 4(3): 1-11.

[32]   Green RE, Krause J, Briggs AW, *et al.* A draft sequence of the Neandertal genome. Science 2010; 328(5979): 710-22.
[http://dx.doi.org/10.1126/science.1188021] [PMID: 20448178]

[33]   Meyer M, Kircher M, Gansauge MT, *et al.* A high-coverage genome sequence from an archaic Denisovan individual. Science 2012; 338(6104): 222-6.
[http://dx.doi.org/10.1126/science.1224344] [PMID: 22936568]

[34]   Henneberg M. Evolution of the human brain: is bigger better? Clin Exp Pharmacol Physiol 1998; 25(9): 745-9.
[http://dx.doi.org/10.1111/j.1440-1681.1998.tb02289.x] [PMID: 9750968]

[35]   Henneberg M, Saniotis A. Evolutionary origins of human brain and spirituality. Anthropol Anz 2009; 67(4): 427-38.
[http://dx.doi.org/10.1127/0003-5548/2009/0032] [PMID: 20440961]

[36]   Henneberg M. The problem of species in hominid evolution. Perspectives in Human Biology 1997; 3: 21-31.

[37]   Quintyn C. The naming of new species in hominin evolution: A radical proposal--A temporary cessation in assigning new names. Homo 2009; 60(4): 307-41.
[http://dx.doi.org/10.1016/j.jchb.2009.05.001] [PMID: 19573870]

[38]   Dobzhansky T, Pavlovsky O. Experimentally created incipient species of Drosophila. Nature 1971; 230(5292): 289-92.
[http://dx.doi.org/10.1038/230289a0] [PMID: 5549403]

[39]   Kluger J. Go fish. (rapid fish speciation in African lakes). Discover 1992; 13(1): 18.

[40]   Henneberg M, Brush G. Similum, a concept of flexible synchronous classification replacing rigid species in evolutionary thinking. Evol Theory 1994; 10: 278.

[41]   De Miguel C, Henneberg M. Variation in hominid brain size: how much is due to method? Homo 2001; 52(1): 3-58.
[http://dx.doi.org/10.1078/0018-442X-00019] [PMID: 11515396]

[42]   Green RE, Krause J, Ptak SE, *et al.* Analysis of one million base pairs of Neanderthal DNA. Nature 2006; 444(7117): 330-6.
[http://dx.doi.org/10.1038/nature05336] [PMID: 17108958]

[43]   Duarte C, MaurA-cio J, Pettitt PB, *et al.* The early Upper Paleolithic human skeleton from the Abrigo do Lagar Velho (Portugal) and modern human emergence in Iberia. Proc Natl Acad Sci USA 1999; 96(13): 7604-9.
[http://dx.doi.org/10.1073/pnas.96.13.7604] [PMID: 10377462]

[44]   Soficaru A, Dobos A, Trinkaus E. Early modern humans from the Pestera Muierii, Baia de Fier,

Romania. Proc Natl Acad Sci USA 2006; 103(46): 17196-201.
[http://dx.doi.org/10.1073/pnas.0608443103] [PMID: 17085588]

[45]　Soficaru A, Petrea C, Dobos A, Trinkaus E. The Human cranium from the Peştera Cioclovina Uscată, Romania: context, age, taphonomy, morphology, and paleopathology. Curr Anthropol 2006; 48(4): 611-9.
[http://dx.doi.org/10.1086/519915]

[46]　Jacob T, Indriati E, Soejono RP, *et al.* Pygmoid Australomelanesian Homo sapiens skeletal remains from Liang Bua, Flores: population affinities and pathological abnormalities. Proc Natl Acad Sci USA 2006; 103(36): 13421-6.
[http://dx.doi.org/10.1073/pnas.0605563103] [PMID: 16938848]

[47]　Eckhardt RB, Henneberg M, Weller AS, HsA1/4 KJ. Rare events in earth history include the LB1 human skeleton from Flores, Indonesia, as a developmental singularity, not a unique taxon. Proc Natl Acad Sci USA 2014; 111(33): 11961-6.
[http://dx.doi.org/10.1073/pnas.1407385111] [PMID: 25092307]

[48]　Henneberg M, Eckhardt RB, Chavanaves S, HsA1/4 KJ. Evolved developmental homeostasis disturbed in LB1 from Flores, Indonesia, denotes Down syndrome and not diagnostic traits of the invalid species Homo floresiensis. Proc Natl Acad Sci USA 2014; 111(33): 11967-72.
[http://dx.doi.org/10.1073/pnas.1407382111] [PMID: 25092311]

[49]　Eckhardt RB. Reply to Westaway. Mandibular misrepresentations fail to support the invalid species *Homo floresiensis*. PNAS 2015. published ahead of print February 6, 2015
[http://dx.doi.org/10.1073/pnas.1422176112]

[50]　Wood B, Collard M. The human genus. Science 1999; 284(5411): 65-71.
[http://dx.doi.org/10.1126/science.284.5411.65] [PMID: 10102822]

[51]　Henneberg M. Brain size/body weight variability in Homo sapiens: Consequences for interpreting hominid evolution. Homo 1990; 39: 121-30.

[52]　Wolpoff MH. Paleoanthropology. 2nd ed., Boston: McGraw Hill 1999.

[53]　Berger LR. Small-bodied humans from Palau, Micronesia. PloS one 2008; 3(3): e1780.

[54]　Henneberg M. Two interpretations of human evolution: essentialism and darwinism. Anthropol Rev 2009; 72: 66-80.

[55]　Hunt KD. The single species hypothesis: truly dead and pushing up bushes, or still twitching and ripe for resuscitation? Hum Biol 2003; 75(4): 485-502.
[http://dx.doi.org/10.1353/hub.2003.0055] [PMID: 14655873]

[56]　Schiaparelli GV. 1893 II Pianeta Marte. new ed., Milano: Vallardi 2002.; Mandrino A, Testa A, Tucci P. Essays in astronomy. New York: D. Appleton and Company 1900.

[57]　Wittgenstein L. Tractatus Logico-Philosophicus. London: Paul Kegan 1922.

[58]　Ereshefsky M. Introduction to The Units of Evolution, Essays on the Nature of Species. Massachusetts: MIT Press XIV 1992.

[59]　Feduccia A. The Origin and Evolution of Birds. 2nd ed., New Haven, Connecticut: Yale University Press 1999.

[60]   Bogin B. The growth of humanity. New York: Wiley Liss 2001; 1: p. 336.

[61]   Oppenheim RW. Cell death during development of the nervous system. Annu Rev Neurosci 1991; 14: 453-501.
[http://dx.doi.org/10.1146/annurev.ne.14.030191.002321] [PMID: 2031577]

[62]   Datta B, Gupta D. The age of menarche in classical India. Ann Hum Biol 1981; 8(4): 351-9.
[http://dx.doi.org/10.1080/03014468100005151] [PMID: 7025746]

[63]   Rurik I, Szigethy E, Fekete F. Relations between anthropometric parameters and sexual activity of Hungarian men. Int J Impot Res 2012; 24(3): 106-9.
[http://dx.doi.org/10.1038/ijir.2011.57] [PMID: 22217602]

[64]   Bell HJ. Consider these factors in cases of suspected infertility 3, 6, 21, 23, 40, 41. Infertility 2012; 61(02): 82.

[65]   Rice WR, Friberg U, Gavrilets S. Homosexuality as a consequence of epigenetically canalized sexual development. Q Rev Biol 2012; 87(4): 343-68.
[http://dx.doi.org/10.1086/668167] [PMID: 23397798]

[66]   Zuber K. Neutrino physics (series in high energy physics, cosmology and gravitation). 2nd ed., USA: Taylor & Francis 2012.

[67]   King RA, Hearing VJ, Creel DJ, Oetting WS. 1995 Albinism. In: Scriver CR, Beaudet AL, Sly WS, Valle D, Eds. The metabolic and molecular basis of inherited disease. 7th ed. USA: McGraw-Hill 1995; pp. 4353-92.

[68]   Nakano KK. Anencephaly: review, developmental medicine & child neurology 1973; 15(3): 1469-8749.

[69]   Birch J. Worldwide prevalence of red-green color deficiency. J Opt Soc Am A Opt Image Sci Vis 2012; 29(3): 313-20.
[http://dx.doi.org/10.1364/JOSAA.29.000313] [PMID: 22472762]

[70]   Jablonski N. Skin, a natural history. Berkeley, Los Angeles: University of California Press 2006.

[71]   Wright MJ, Irving MD. Clinical management of achondroplasia. Arch Dis Child 2012; 97(2): 129-34.
[http://dx.doi.org/10.1136/adc.2010.189092] [PMID: 21460402]

[72]   Vakoch DA, Ed. Astrobiology, history and society. Heidelberg: Springer 2013.
[http://dx.doi.org/10.1007/978-3-642-35983-5]

[73]   Tollervey JR, Lunyak VV. Epigenetics: judge, jury and executioner of stem cell fate. Epigenetics 2012; 7(8): 823-40.
[http://dx.doi.org/10.4161/epi.21141] [PMID: 22805743]

[74]   Macleod EL, Ney DM. Nutritional management of phenylketonuria. Ann Nestle [Eng] 2010; 68(2): 58-69.
[http://dx.doi.org/10.1159/000312813] [PMID: 22475869]

[75]   Komlos J, Ed. Stature, living standards and economic history. 1st ed., Chicago: University of Chicago Press 1994.

[76]   Henneberg M. Continuing human evolution: brains, bodies, and the role of variability. Trans R Soc S

Afr 1992; 48(1): 159-82.
[http://dx.doi.org/10.1080/00359199209520260]

[77]     Henneberg M. The llusive concept of human variation: thirty years of teaching biological anthropology on four continents. In: Strkalj G, Ed. Teaching human variation: trends, issues and challenges. Nova Science Publishers, Inc. 2010; pp. 34-41.

[78]     Henneberg RJ, Henneberg M. Variation in the closure of the sacral canal in the skeletal sample from Pompeii, Italy 79 AD. Hum Biol 1999; 4(1): 177-88.

[79]     Solomon LB, RA1/4hli FJ, Lee YC, Henneberg M. Secular trend in the opening of the sacral canal: an Australian study. Spine 2009; 34(3): 244-8.
[http://dx.doi.org/10.1097/BRS.0b013e3181908ca2] [PMID: 19179919]

[80]     Lee YC, Solomon LB, RA1/4hli FJ, *et al.* Confirmation of microevolutionary increase in spina bifida occulta among Swiss birth cohorts. Eur Spine J 2011; 20(5): 776-80.
[http://dx.doi.org/10.1007/s00586-010-1519-2] [PMID: 20632043]

[81]     Henneberg M, George BJ. Possible secular trend in the incidence of an anatomical variant: median artery of the forearm. Am J Phys Anthropol 1995; 96(4): 329-34.
[http://dx.doi.org/10.1002/ajpa.1330960402] [PMID: 7604889]

[82]     Bhatia K, Ghabriel M, Henneberg M. Anatomical variations in the branches of the human arch of the aorta: a possible increase in recent times? Folia Morphol (Warsz) 2005; 64(3): 217-23.
[PMID: 16228958]

[83]     Henneberg M, Steyn M. Trends in cranial capacity and cranial index in Subsaharan Africa during the Holocene. Am J Hum Biol 1993; 5: 473-9.
[http://dx.doi.org/10.1002/ajhb.1310050411]

[84]     Mathers K, Henneberg M. Were we ever that big? gradual increase in hominid body size over time. Homo 1995; 46: 141-73.

[85]     Henneberg M, Henneberg RJ. Reconstructing medical knowledge in ancient Pompeii from the hard evidence of bones and teeth. Roma: "L'Erma" di Bretschneider 2002.

[86]     Henneberg M, Veitch D. National size and shape survey of Australia. Kinanthreport 2003; 16: 34-9.; Henneberg M, Veitch D. Is obesity as measured by body mass index, and waist circumference in adult Australian women 2002 just a result of the lifestyle? J Human cology Special issue 2003; 13: 85-9.

[87]     Henneberg M. Secular trends in body height – indicator of general improvement in living conditions or of a change in specific factors? In: Dasgupta P, Hauspie R, Eds. Perspectives in human growth, development and maturation. Boston: Kluwer Academic Publishers 2001; pp. 159-68.
[http://dx.doi.org/10.1007/978-94-015-9801-9_14]

[88]     Ehrlich P. Human natures: genes, culture and the human prospect. United States: Penguin Books 2000.

[89]     Carson R. Silent spring. Boston: Houghton Mifflin Co 1962.

[90]     Jenkins M. The dystopian world of blade runner: an ecofeminist perspective. Trumpeter 1997; 14(4): 1-12.

[91]     Tucker ME. Thomas Berry and the new Story: an introduction to the work of Thomas Berry. In: Eaton H, Ed. The intellectual journey of Thomas Berry: imagining the Earth community. Lanham, MD:

Lexington Books 2014.

[92]  Eliade M. Patterns in comparative religion. Cleveland: The World Publishing Company 1963.

[93]  Spencer C. Extract from the heretics feast: a history of vegetarianism. London: University of New England 1995.

[94]  Wei-Ming T. Humanity and self cultivation: essays in Confucian thought. Berkeley: Asian Humanities Press 1979.

[95]  Nasr SH. Introduction, science and civilization in Islam. New York: New American Library 1968.

[96]  Hope M, Young J. Islam and ecology. Crosscurrent 1994; 44(2): 3.

[97]  Rodman J. "Four forms of ecological consciousness reconsidered," deep ecology for the 21st century. Boston: Shambhala Press 1995.

[98]  Antolick M. Deep ecology and Heidegerrian phenomenology MA thesis. Florida, USA: Department of Philosophy. University of South Florida 2002; p. 56.

[99]  Foucault M. The birth of the clinic: an archeaology of medical perception. New York: Vintage Books 1975.

[100]  Said E. Orientalism. New York: Vintage 1979.

[101]  Orientalism. http://www.english.emory.edu/Bahri/Orientalism.html

[102]  James SP. Heidegger and the role of the body in environmental values. The Trumpeter. 2002; 18.(1)

[103]  Weber M. From Max Weber: essays in sociology. New York: Oxford University Press 1946.

[104]  Angyal AJ. Thomas Barry's Spirituality and the "Great Work". Ecozoic Reader 2003; 3(3): 35-44.

[105]  Kinsley D. Ecology and religion: ecological spirituality in cross-cultural perspective. Englewood Cliffs, NJ: Prentice-Hall 1995.

[106]  Abram D. The ecology of magic. Chapter excerpt from Spell of the Sensuous: Perception and Language in a More-Than-Human World. 1997.

[107]  Berry T. The dream of the earth. San Francisco: Sierra Club Books 1988.

[108]  Berry T. Earth spirituality. Riverdale Papers. New York: the Riverdale Center for Religious Studies, n. d. 1990; V.

[109]  Berry J. The universe story, As told by Brian Swimme and Thomas Berry. Trumpeter 1993; 10(2)

[110]  Swimme B. A scientist responds to Thomas Berry's cosmology. The Ecozoic Reader 2003; 3(4): 23.

[111]  Berry T. The spirituality of the Earth. In: Birch C, Eaken W, McDaniel , Eds. Liberating life: contemporary approaches in ecological theology. Maryknoll, New York: Orbis Books 1990; pp. 151-8.

[112]  Swimme B. The universe is a green dragon: reading the meaning in the cosmic story. A Quarterly of Humane Sustainable Culture. 1997; pp. 1-14.

[113]  Linde A. A new inflationary universe scenario: a possible solution of the horizon, flatness, homogeneity, isotropy, and primordial monopole problems. Phys Lett 1982; 108B: 389-92. [http://dx.doi.org/10.1016/0370-2693(82)91219-9]

[114]  Kazanas D. Dynamics of the universe and spontaneous symmetry breaking. Astrophys J 1980; 241:

L59-63.
[http://dx.doi.org/10.1086/183361]

[115]   Swimme B, Tucker ME. Journey of the universe. New Haven: Yale University Press 2011.

[116]   Davies P. God and the New Physics. New York: Simon & Schuster 1984.

[117]   What is the "fine-tuning" of the universe, and how does it serve as a "pointer to God"? The Biologos Forum: Science and Faith in Dialogue. 2015.

[118]   Stenger V. Is the Universe Fine-Tuned for Us? In: Young M, Edis T, Eds. Why intelligent design fails: a scientific critique of the new creationism. New Jersey, USA: Rutgers University Press 2004; pp. 172-84.

[119]   Linde A. Particle physics and inflationary cosmology. New York: Academic Press 1990; pp. 1-90.
[http://dx.doi.org/10.1016/B978-0-12-450145-4.50004-9]

[120]   Linde A. The self-reproducing inflationary universe. Sci Am 1994; 271: 48-55.
[http://dx.doi.org/10.1038/scientificamerican1194-48]

[121]   Johan R, *et al.* Discovery of a possibly old galaxy at z=6.027, multiply imaged by the massive cluster Abell 383. Monthly Notices of the Royal Astronomical Society 2011.
[http://dx.doi.org/10.1111/j.1745-3933.2011.01050.x]

[122]   Yoshida N, Omukai K, Hernquist L. Protostar formation in the early universe. Science 2008; 321(5889): 669-71.
[http://dx.doi.org/10.1126/science.1160259] [PMID: 18669856]

[123]   Kambiz F. Revisiting the scale length–µ0 plane and the freeman Law in the local universe. Astrophys J 2010; 722(1): L120.
[http://dx.doi.org/10.1088/2041-8205/722/1/L120]

[124]   Lotz JM, Jonsson Patrik, Cox TJ, *et al.* The major and minor galaxy merger rates at z < 1.5. Astrophys J 2011; 742(103)
[http://dx.doi.org/10.1088/0004-637X/742/2/103]

[125]   Stott JP, Collins CA, Burke C, Hamilton-Morris V, Smith GP. Little change in the sizes of the most massive galaxies since z = 1. Mon Not R Astron Soc 2011.
[http://dx.doi.org/10.1111/j.1365-2966.2011.18404.x]

[126]   Swimme B, Thomas B. The universe story. New York, USA: Harper One 1992.

[127]   Hoyle F, Wickramasinghe C, Watson J. Viruses from Space and Related Matters . UK: University College Cardiff Press 1986.

[128]   NASA researchers. DNA building blocks can be made in space. NASA 2011.

[129]   NASA Researchers. Make First Discovery of Life's Building Block in Comet NASA 2009.

[130]   Wickramasinghe NC, Wainwright M, Narlikar JV, Rajaratnam P, Harris MJ, Lloyd D. Progress towards the vindication of panspermia. Astrophys Space Sci 2003; 283: 403-13.
[http://dx.doi.org/10.1023/A:1021677122937]

[131]   Vaidya PG. Critique on vindication of panspermia. Apeiron 2009; 15(3): 463-74.

[132]   Becoquerel P. La suspension de la vie au dessous de 1/20 absolu par demagnetization adiabatique de

l'alun de fer dans le vide les plus eléve. C.R. Hebd Séances Acad Sci Paris 1950; 231: 261-3.

[133] JAnsson KI, Rabbow E, Schill RO, Harms-Ringdahl M, Rettberg P. Tardigrades survive exposure to space in low Earth orbit. Curr Biol 2008; 18(17): R729-31.
[http://dx.doi.org/10.1016/j.cub.2008.06.048] [PMID: 18786368]

[134] Jönsson KI. Tardigrades as a potential model organism in space research. 2007; 7: 757-766.29-R731.
[129] Gladyshev E, Meselson M. . Astrobiology 2007; 757-66.

[135] Gladyshev E, Meselson M. Extreme resistance of bdelloid rotifers to ionizing radiation. Proc Nat Acad Sci 2008; 5: 5139-44.

[136] Cano RJ, Borucki MK. Revival and identification of bacterial spores in 25- to 40-million-year-old Dominican amber 1995; 268: 1060-4.

[137] Schidlowski MA. 3,800-million-year isotopic record of life from carbon in sedimentary rocks. Nature 1988; 333: 313-8.
[http://dx.doi.org/10.1038/333313a0]

[138] Margulis L. Interviewed in The End of Science, by John Horgan. Addison-Wesley Publishing Company, Inc 1999; pp. 140-1.

[139] Klyce Brig. Cosmic Ancestry website. http://www.panspermia.org/abouthis.htm

[140] Moore PB. Ribosomes and the RNA world. In: Gesteland RF, Atkins JF, Eds. The RNA world. Cold Spring Harbor: Laboratory Press 1993; pp. 119-35.

[141] FranAois J. Of flies mice and men. Cambridge, MA: Harvard University Press 1998.

[142] Winkelman M. Shamanism as the original neurotheology. Zygon 2004; 39(1): 193-217.
[http://dx.doi.org/10.1111/j.1467-9744.2004.00566.x]

[143] Winkelman M. Shamanism: the neural ecology of consciousness and healing Westport, Connecticut. London: Bergin & Garvey 2000.

[144] Winkelman M. Shamanisn and cognitive evolution. Camb Archaeol J 2002; 12(1): 71-101.
[http://dx.doi.org/10.1017/S0959774302000045]

[145] Saniotis A, Henneberg M. Manifestations of mystical experience and evolution of the human brain. Hum Evol 2011; 26(1-2): 61-74.

[146] Saniotis A, Henneberg M. Explorations into the biology of emotion and religious experience. Int J Anthropol 2011; 26(1-2): 25-36.

[147] McClenon J. Shamanic healing, human evolution, and the origin of religion. J Sci Study Relig 1997; 36(3): 345-54.
[http://dx.doi.org/10.2307/1387852]

[148] Leroi-Gourhan A. Treasures of prehistoric art. New York: Harry N. Abrams 1967.

[149] Ryan R. The strong eye of shamanism: a journey into the caves of consciousness. Rochester, New York: Inner Traditions 1999.

[150] Eliade M. The two and the one. New York: Harper Torchbooks 1965.

[151] Michell J. City of revelation: on the proportion and symbolic numbers of the cosmic temple. London: Abacus 1973.

[152] Abram D. Spell of the sensuous: perception and language in a more-than-human world. New York: Vintage Books 1997.

[153] Sepharial K. Ancient secrets of numerology. New Delhi: Orient Paperbacks 1994.

[154] Bond FB, Lea TS. Gematria: a preliminary investigation of the Cabala. London: Thorsons Publishers Limited 1977.

[155] Riedweg C. Pythagoras: his life, teaching and influence. Cornell: Cornell University Press 2005.

[156] White M. Isaac Newton: The last sorcerer. Reading, MA: Addison-Wesley 1997.

[157] Stapleton G. Probable source of numbers on which Jabrian alchemy was based. Arch Intern Hist Sci 1953.

[158] Saggs HW. The greatness that was Babylon: a sketch of the ancient civilization of the Tigris-Euphrates alley. New York: Mentor Books 1962.

[159] Jackson M. Minica ethnographica: intersubjectivity and the anthropological project. London: The University of Chicago Press 1998.

[160] Van Gennep A. The rites of passage. London: Routledge 2004. (Original work published 1960)

[161] Turner V. The forest of symbols: aspects of Ndembu ritual. Ithaca, N.Y: Cornell University Press 1967.

[162] Turner V. The drums of affliction: a study of religious processes among the Ndembu of Zambia Oxford: Clarendon Press. London: International African Institute 1968.

[163] Turner V. The ritual process: structure and anti-structure. Chicago: Aldine 1969.

[164] Turner V. Encounter with Freud: the making of a comparative symbologist. In: Spindler G, Spindler L, Eds. The making of psychological anthropology. Berkeley: University of California Press 1978; pp. 558-83.

[165] Buber M. I and thou. New York: Charles Scribner's Sons 1958.

[166] Christopher JB. The Islamic tradition. New York: Harper & Row 1976.

[167] Akkach S. In the image of the cosmos order and symbolism in traditional Islamic architecture Part (2). Islam Q 1995; 39(2): 93.

[168] Turner VW. Dramas, fields, and metaphors: symbolic action in human society. Ithaca: Cornell University Press 1974.

[169] Meyerhoff B. Peyote hunt: the sacred journey of the Huichol indians. New York: Cornell University Press 1976.

[170] Eliade M. Myths, dreams and mysteries Phillip Mairet (trans). London: Harvell Press 1960.

[171] Otto R. The idea of the holy Translated by JW. New York: Oxford 1958.

[172] Bloom H. Global brain: the evolution of mass mind from the big bang to the 21st Century. New York: John Wiley & Sons, Inc 2000.

[173] Teilhard de Chardin P. The phenomenon of man. New York: Harper Perennial 1976. (Original work published 1959)

[174]  Gill JH. Merleau-Ponty and metaphor. New Jersey: Humanities Press 1991.

[175]  Izutsu T. Sufism and Taoism: a comparative study of the key philosophical concepts. Tokyo: Iwanami Shoten 1983.

[176]  Bourguignon E. World distribution and patterns of possession states. In: Prince R, Ed. Trance and possession states. Montreal: R. M. Bucke Memorial Society 1968; pp. 3-34.

[177]  Charlesworth M. Science, non-science & pseudo-Science. Burwood, Victoria: Deakin University Press 1982.

[178]  Dewey J. Experience and nature. New York: Dover 1958.

[179]  Vandana S. Monocultures of the mind. Trumpeter 1993; 10(4); Kuhn T. The structure of scientific revolutions. Chicago: University of Chicago Press 1972.

[180]  Horton R. African traditional thought and western science. Africa: J Int African Inst 1967; 37(2): 157-87.

[181]  Radin D. Testing nonlocal observation as a source of intuitive knowledge. Explore (NY) 2008; 4(1): 25-35.
[http://dx.doi.org/10.1016/j.explore.2007.11.001] [PMID: 18194788]

[182]  Winkelman M. Shamanism as neurotheology and evolutionary psychology. Am Behav Sci 2002; 45(12): 1873-85.
[http://dx.doi.org/10.1177/0002764202045012010]

[183]  Laughlin CD. The mystical brain: biogenetic structural studies in the anthropology of religion. Available at: http://www.biogeneticstructuralism.com/articles.htm. 1996.

[184]  Uzendoski M. Somatic poetry in Amazonian Ecuador. Anthropol Humanism 2008; 33(1-2): 12-29.
[http://dx.doi.org/10.1111/j.1548-1409.2008.00002.x]

[185]  Radin D. Testing nonlocal observation as a source of intuitive knowledge. Explore (NY) 2008; 4(1): 25-35.
[http://dx.doi.org/10.1016/j.explore.2007.11.001] [PMID: 18194788]

[186]  Tedlock B. Toward a theory of divinatory practice. Anthropol Consciousness 2006; 17(2): 62-77.
[http://dx.doi.org/10.1525/ac.2006.17.2.62]

[187]  Cosmides L, Tooby J. Are humans good intuitive statisticians after all? Rethinking some conclusions from the literature on judgment under uncertainty. Cognition 1996; 58: 1-73.
[http://dx.doi.org/10.1016/0010-0277(95)00664-8]

[188]  Newberg A, D'Aquilli E, Rause V. Why God won't go away. New York: Ballantine Books 2002.

[189]  Bem DJ. Feeling the future: experimental evidence for anomalous retroactive influences on cognition and affect. J Pers Soc Psychol 2011; 100(3): 407-25.
[http://dx.doi.org/10.1037/a0021524] [PMID: 21280961]

[190]  Schutz A, Luckmann T. The structure of the life world. Illinois: Northwestern University Press 1973.

[191]  Levesque-Lopman L. Claiming reality: phenomenology and women's experience. New Jersey: Rowman & Littlefield 1988.

[192] Nuckolls CW. Deciding how to decide: possession-mediumship in Jalari divination. Med Anthropol 1991; 13(1-2): 57-82.
[http://dx.doi.org/10.1080/01459740.1991.9966041] [PMID: 1881300]

[193] Wilce JM. Divining TROUBLES or divining troubles? Gender, conflict, and polysemy in Bangladeshi divination. Anthropol Q 2001; 74(4): 190-9.
[http://dx.doi.org/10.1353/anq.2001.0040]

[194] Myhre KC. Divination and experience: explorations of a Chagga epistemology. J R Anthropol Inst 2006; 12(2): 313-30.
[http://dx.doi.org/10.1111/j.1467-9655.2006.00293.x]

[195] Eglash R. Bamana sand divination: recursion in ethnomathematics. Am Anthropol 1997; 99(1): 112-22.
[http://dx.doi.org/10.1525/aa.1997.99.1.112]

[196] Schlitz M, Braud W. Distant intentionality and healing: assessing the evidence. Altern Ther Health Med 1997; 3(6): 62-73.
[PMID: 9375431]

[197] Grad B. Some biological effects of laying-on of hands: a review of experiments with animals and plants. J Am Soc Psych Res 1965; 59a: 95-127.

[198] Grad B, Cadaret RJ, Paul GI. The influence of an unorthodox method of treatment of wound healing in mice. Int J Parapsychol 1961; 3: 5-24.

[199] Chen KW, Shiflett SC, Ponzio NM, He B, Elliott DK, Keller SE. A preliminary study of the effect of external qigong on lymphoma growth in mice. J Altern Complement Med 2002; 8(5): 615-21.
[http://dx.doi.org/10.1089/107555302320825138] [PMID: 12470443]

[200] Radin D, Yount G, Yount G. Effects of healing intention on cultured cells and truly random events. J Altern Complement Med 2004; 10(1): 103-12.
[http://dx.doi.org/10.1089/107555304322849020] [PMID: 15029876]

[201] Sicher F, Targ E, Moore D, Smith HS. A randomized double-blind study of the effect of distant healing in a population with advanced AIDS. Report of a small scale study. West J Med 1998; 169(6): 356-63.

[202] Laszlo E. Science and the akhashic field: an integral theory of everything. Rochester, Vermont: Inner Traditions 2004.

[203] Combs A, Arcari T, Krippner S. All of the myriad worlds: life in the Akashic plenum. World Futures 2006; 62: 75-85.
[http://dx.doi.org/10.1080/02604020500412709]

[204] Bache CM. Reincarnation and the akhashic field: a dialogue with Ervin Laszlo. World Futures 2006; 62: 114-26.
[http://dx.doi.org/10.1080/02604020500412873]

[205] Zyga L. Physicists calculate number of parallel universes. PhysOrg. http://phys.org/news/2009-10-physicists-parallel-universes.html 2009.

[206] de Quincey C. The "metaverse story": where science meets spirit. In: Laszlo E, Ed. Science and the

reenchantment of the cosmos. Rochester, Vermont: Inner Traditions 2006; pp. 109-20.

[207] AndreA' LG. L'Art Parie'tal: langage de la pre'histoire . Grenoble: Millon 1992.

[208] Lewis-Williams D. Believing and seeing: symbolic meanings in Southern San rock paintings. London: Academic Press 1981.

[209] Power C. Women in prehistoric rock art. In: Berghaus G, Ed. New perspectives on prehistoric art. Westport, CT/London: Praeger 2004; pp. 75-103.

[210] Lidman MJ. Werewolves in psyche and cinema: man-beast transformation and paradox. J Pop Cult 2004; 10(2): 388-97.

[211] Stewart TT. The origin of the werewolf superstition University of Missouri Studies, Social Science Series, 2 (3). Columbia: University of Missouri Press 1909; 2.(3)

[212] Weiss D. It's a dog's life. Discovery: Channel Magazine 2010; 3: 14.

[213] White DG. Myths of the dog-an. Chicago: The University of Chicago Press 1991.

[214] Trut L. Early canid domestication: the farm-fox experiment. Am Sci 1999; 87(2): 160.
[http://dx.doi.org/10.1511/1999.2.160]

[215] Balzer MM. Flights of the sacred: symbolism and theory in Siberian shamanism. Am Anthropol 1996; 98(2): 305-18.
[http://dx.doi.org/10.1525/aa.1996.98.2.02a00070]

[216] Popov AA. Poluchenie 'shamanskogo dara' u Villiuiskikh Yakutov. (Receiving the 'shamanic gift' among the Villiuisk Yakut). Trudy Instituta Etnografi Akademii Nauk 1947; 2: 282-93.

[217] Abram D. Becoming Human: An Earthly Cosmology. New York: Vintage Books 2010.

[218] Saniotis A. Enchanted landscapes: sensuous awareness as mystical practice among north indian sufis. Aust J Anthropol 2008; 19(1): 17-26.
[http://dx.doi.org/10.1111/j.1835-9310.2008.tb00103.x]

[219] Jackson M. Allegories of the wilderness: ethics and ambiguity in Kuranko narratives. Bloomington: Indiana University Press 1982.

[220] Taussig M. Mimesis and alterity: a particular history of the senses. London, New York: Routledge 1993.

[221] Michell J. How the world is made: the story of creation according to sacred geometry. London: Thames & Hudson 2012.

[222] Margulis L. Symbiotic planet: a new look at evolution. New York: Basic Books 1998.

[223] Keeney B. Circular epistemology and the bushman shamans: a Kalahari challenge to the hegemony of narrative. Cybern Hum Knowing 1998; 12(1-2): 75-89.

[224] Kurzweil R. The age of spiritual machines: when computers exceed human intelligence. New York: Penguin Books 2000.

[225] Drexler KE. Nanosystems: molecular machinery, manufacturing, and computation. New York: John Wiley & Sons, Inc 1992.; Bostrom N. Are you living in a computer simulation? Philos Q 2003; 53(211): 243-55.

[226]  Lloyd S. Ultimate physical limits to computation. Nature 2000; 406(6799): 1047-54. [http://dx.doi.org/10.1038/35023282] [PMID: 10984064]

[227]  Giacomo PDe, Mich L, Storelli M, *et al.* A method to increase students' cognitive potentialities using the elementary pragmatic model. In: Macer DRJ, Ed. Challenge for bioethics From asia: the behaviourome project. 2004; pp. 95-101.

[228]  Bostrom N. The World in 2050. Broadcast by BBC Virtual Reality. Available at: http://www.nickbostrom.com/2050/world.html 2000.

[229]  Bostrom N. What is Transhumanism? Available at: http://www.nick-bostrom.com/old/transhumanism.html 2001. (Original version appeared in 1998, here slightly revised and with a postscript added in 2001)

[230]  Tipler FJ. The physics of immortality: modern cosmology, god and the resurrection of the dead. London: Macmillan 1995.

[231]  Allhoff F. Germ-line genetic enhancement and Rawlsian primary goods. J Evol Technol 2008; 18(1): 10-26.

[232]  Greenfield S. Tomorrow's people: how 21st century technology is changing the way we think and feel. London: Penguin Books 2003; p. 304.

[233]  Krueger O. Gnosis in cyberspace. J Evol Technol 2005; 14(2): 55-67.

[234]  Moravic H, Pohl F. Souls in silicon. Omni 1993; 11: 66-76.

[235]  De Hallowell AI. The role of dreams in Ojibwa culture. In: von Ginnebaum GE, Collins R, Eds. The Dream in Human Societies. Berkeley: University of California Press 1966; pp. 287-92.

[236]  Merrill N. The Ratamuri stereotype of dreams. In: Tedlock B, Ed. Dreaming: anthropological and psychological approaches. New York: Cambridge University Press 1987; pp. 194-219.

[237]  Shaw R. Dreaming as accomplishment: power, the individual, and Tenne divination. In: Jedrej MC, Shaw R, Eds. Dreaming, religion and society in Africa. Leiden: E.J. Brill 1992; pp. 36-54.

[238]  Tuzin D. The breath of a ghost: dreams and fear of the dead. Ethos 1975; 2: 555-78. [http://dx.doi.org/10.1525/eth.1975.3.4.02a00050]

[239]  Worthman CM, Melby M. Toward a comparative developmental ecology of human sleep. In: Carskadon MA, Ed. Adolescent sleep patterns: biological, social, and psychological influences. New York: Cambridge University Press 2002; pp. 69-117.

[240]  Price-Williams D, Degarrod LN. Dreams as interaction. Anthropol Consciousness 1996; 7(2): 16-23. [http://dx.doi.org/10.1525/ac.1996.7.2.16]

[241]  Saniotis A. Encounters with the religious imagination and the emergence of creativity. World Futures. J General Evolution 2009; 65(7): 464-76.

[242]  Desjarlais RR. Body and emotion: the aesthetics of illness and healing in the Nepal Himalayas. Delhi: Motilal Banarsidass Publishers 1994.

[243]  Stone MR. Creativity in dreams. http://www.dreamresearch.ca/pdf/creativity.pdf

[244]  Baaquie BE, Willeboordse FH. Exploring integrated science. Florida: Taylor and Francis Group 2010.

[245] Linn D. The hidden power of dreams: the mysterious world of dreams revealed. USA: Hay House 2009.

[246] Dilley R. Dreams, insipiration, and craftwork among Tukular weavers. In: Jedrej MC, Shaw R, Eds. Dreaming, religion and society in Africa. Leiden: E.J. Brill 1992; pp. 71-85.

[247] Glaskin K. Dreams, memory, and the ancestors: creativity, culture, and the science of sleep. J R Anthropol Inst 2011; 17(1): 44-62.
[http://dx.doi.org/10.1111/j.1467-9655.2010.01668.x]

[248] Devereux G. Dream learning and individual ritual differences in Mohave shamanism. Am Anthropol 1957; 59: 1036-45.
[http://dx.doi.org/10.1525/aa.1957.59.6.02a00080]

[249] Eves R. Pentecostal dreaming and technologies of governmentality in a Melanesian society. Am Ethnologist 2011; 38(4): 758-73.
[http://dx.doi.org/10.1111/j.1548-1425.2011.01335.x]

[250] Bulkeley K. Dreaming as a spiritual practice. Anthropol Consciousness 1996; 7(2): 1-15.
[http://dx.doi.org/10.1525/ac.1996.7.2.1]

[251] Trafzer CE, Beach MA. Smohalla, the Washani, and religion as a factor in Northwestern Indian history. Am Indian Q 1987; 10(3): 217-41.

[252] Franklin MS. The role of dreams in the evolution of the human mind. Evol Psychol 2005; 3: 59-78.
[http://dx.doi.org/10.1177/147470490500300106]

[253] Hall C, Van de Castle R. The content analysis of dreams. New York: Appleton-Century-Crofts 1966.

[254] University of the West of England. Women have more nightmares than men, study shows. ScienceDaily 2009. http://www.sciencedaily.com/releases/2009/01/090128104535.htm

[255] Brizendine L. The female brain. London: Bantam Books 2007.

[256] American Academy of Sleep Medicine. Dream-enacting behavior is common in healthy young adults. ScienceDaily 2012. http://www.sciencedaily.com/releases/2009/12/091201084049.htm

[257] van der Helm E, Yao J, Dutt S, Rao V, Saletin JM, Walker MP. REM sleep depotentiates amygdala activity to previous emotional experiences. Curr Biol 2011; 21(23): 2029-32.
[http://dx.doi.org/10.1016/j.cub.2011.10.052] [PMID: 22119526]

[258] Domholl GW. The scientific study of dreams: neural networks, cognitive developments, and content analysis. Washington, D.C.: American Psychological Association 2003.

[259] Stephens M. Self, the sacred other, and autonomous imagination. In: Herdt G, Stephen M, Eds. The religious imagination in New Guinea. London: Rutgers University Press 1989; pp. 41-64.

[260] Hobson JA. 13 Dreams Freud never had: the new mind science. New York: Pi Press 2005.

[261] Dement WC. Some must watch while some must sleep. San Francisco: San Francisco Book Co 1972.

[262] University of California - San Diego. Let me sleep on it: creative problem solving enhanced by REM sleep. ScienceDaily 2012. http://www.sciencedaily.com/releases/2009/06/090608182421.htm

[263] Ichikawa J. Dreaming and imagination. Mind Lang 2009; 24(1): 103-21.
[http://dx.doi.org/10.1111/j.1468-0017.2008.01355.x]

[264] Pearce T. Kids have been getting less sleep than recommended for a century. The Globe and Mail. Available at: http://www.theglobeandmail.com/life/the-hot-button/kids-have-been-getting-less-sleep-than-recommended-for-a-century/article621251/. 2012.

[265] Dodds ER. The Greeks and the irrational. Berkeley: University of California Press 1951.

[266] Herdt G, Stephen M, Eds. The religious imagination in New Guinea. London: Rutgers University Press 1989; p. 19.

[267] Vernant JP. Myth and thought Among the Greeks. London: Routledge & Kegan Paul 1983.

[268] Plato . Phaedrus and the seventh and eighth Letters. Middlesex, England: Penguin 1974.

[269] Phillips AA. Ibn Taymeeyah's essay on the jinn (demons). Riyadh, Saudi Arabia: Tawheed Publications 1989.

[270] Mylonas GE. Mycenae and the Mycenaean Age. Princeton, N.J.: Princeton University Press 1966.

[271] Nilsson M. The Greek popular religion. New York: Columbia University Press 1940.

[272] Nilsson M. Die Geschichte der Griechische religion. Munchen: C.F.Beck Verlag 1967; I.

[273] Kerényi K. Dionysos: archetypal image of indestructible life. Princeton: Bollingen 1976.

[274] McEvilley T. The shape of ancient thought. New York: Allsworth Press 2002.

[275] Ruden S. Hymns: To Dionysus' (1, 7 and 26). In: Ruden S, Ed. Homeric hymns. Indianapolis: Hackett 2005.

[276] Morford MO, Lenardon RJ. Classical Mythology 5. New York: Longman 1995.

[277] Broad WJ. The oracle: ancient Delphi and the science behind its lost secrets. New York: Penguin Press 2007.

[278] Kerényi K. Eleusis: archetypal image of mother and daughter. Princeton, N.J.: Princeton University Press 1991.

[279] Cline K. The shaman's song and divination in the epic tradition. Anthropol Consciousness 2010; 21(2): 163-87.
[http://dx.doi.org/10.1111/j.1556-3537.2010.01027.x]

[280] Wright SH. Childhood influences that heighten psychic powers. J Spiritual Paranormal Stud 2006; 29(4): 183-93.

[281] Cohn SA. Second sight and family history: pedigree and segregation analyses. J Sci Explor 1999; 13(3): 351-72.

[282] Cohn SA. A questionnaire study on second sight experiences. J Soc Psychical Res 1999; 63(855): 129-57.

[283] Sidgwick H, Johnson A, Myers FWH, Podmore F, Sidgwick EM. Report on the census of hallucinations. Proceedings of the Society for Psychical Research. Forgotten Books 1894; 10: pp. 25-422.

[284] Radford T. Men guessed right on women's intuition. The Guardian 2005.

[285] Why Are Some Paranormal Beliefs More Attractive to Males While Others Are More Appealing to

Females? http://www.scienceandreligiontoday.com/2011/02/18/why-are-some-paranormal-beliefs-more-attractive-to-males-while-others-are-more-appealing-to-females/ 2011.

[286] Jung CG. Man and symbols. London: Aldous Books 1964.

[287] University of British Columbia. Analytic thinking can decrease religious belief, study shows ScienceDaily 2012. http://www.sciencedaily.com/releases/2012/04/120426143856.htm

[288] Bostrom N, Sandberg A. Cognitive enhancement: methods, ethics, regulatory challenges. Oxford: Oxford University. The Future of Humanity Institute 2006.

[289] Pieters T, Snelders S. Psychotropic drug use: between healing and enhancing the mind. Neuroethics 2009; 2(2): 1-11.
[http://dx.doi.org/10.1007/s12152-009-9033-0]

[290] Farah MJ, Illes J, Cook-Deegan R, *et al.* Neurocognitive enhancement: what can we do and what should we do? Nat Rev Neurosci 2004; 5(5): 421-5.
[http://dx.doi.org/10.1038/nrn1390] [PMID: 15100724]

[291] Naam R. More than human: enhancing the promise of biological enhancement. New York: Broadway 2005.

[292] Miller KR. Finding Darwin's god: a scientist's search for common ground between god and evolution. New York: Cliff Street Books 1999.

[293] Shermer M. Science Friction: Where the Known Meets the Unknown. New York: Henry Holt/Times Books 2005.

[294] Linden DL. The Accidental mind: how brain evolution has given us love, memory, dreams, and god. 1 ed. Cambridge, Mass: Harvard University Press 2008; p. 276.

[295] Horner V, Whiten A. Causal knowledge and imitation/emulation switching in chimpanzees (Pan troglodytes) and children (Homo sapiens). Anim Cogn 2005; 8(3): 164-81.
[http://dx.doi.org/10.1007/s10071-004-0239-6] [PMID: 15549502]

[296] Sherry D, Duff S. Behavioural and neural bases of orientation in food-storing birds. J Exp Biol 1996; 199(Pt 1): 165-72.
[PMID: 9317563]

[297] Inoue S, Matsuzawa T. Working memory of numerals in chimpanzees. Curr Biol 2007; 17(23): R1004-5.
[http://dx.doi.org/10.1016/j.cub.2007.10.027] [PMID: 18054758]

[298] Marino L. Cetaceans and primates: Convergence in intelligence and self-awareness J Cosmology 2011; 14

[299] Thacker E. Biomedia. Minneapolis & London: Minnesota 2004; p. 240.

[300] Nicolelis MA, Srinivasan M. Human-machine interactions: potential impact of nanotechnology in the design of neuroprosthetic devices aimed at restoring or augmenting human performance. In: Roco MC, Bainbridge WS, Eds. Converging technologies for improving human Performance: nanotechnology, biotechnology, information technology and cognitive science. Dordrecht, The Netherlands: Kluwer Academic Press 2003; pp. 251-5.

[301] Hoag H. Remote control. Nature 2003; 423(6942): 796-8.

[http://dx.doi.org/10.1038/423796a] [PMID: 12815397]

[302] Andersen RA, Burdick JW, Musallam S, Pesaran B, Cham JG. Cognitive neural prosthetics. Trends Cogn Sci (Regul Ed) 2004; 8(11): 486-93.
[http://dx.doi.org/10.1016/j.tics.2004.09.009] [PMID: 15491902]

[303] Foster KR. Engineering the brain. In: Iles J, Ed. Neuroethics: defining the issues in theory, practice, and policy. Oxford: Oxford University Press 2006; pp. 185-99.

[304] Sanchez JC, Carmena JM, Lebedev MA, Nicolelis MA, Harris JG, Principe JC. Ascertaining the importance of neurons to develop better brain-machine interfaces. IEEE Trans Biomed Eng 2004; 51(6): 943-53.
[http://dx.doi.org/10.1109/TBME.2004.827061] [PMID: 15188862]

[305] Kim SP, Sanchez JC, Erdogmus D, *et al.* Divide-and-conquer approach for brain machine interfaces: nonlinear mixture of competitive linear models. Neural Netw 2003; 16(5-6): 865-71.
[http://dx.doi.org/10.1016/S0893-6080(03)00108-4] [PMID: 12850045]

[306] Friehs GM, Zerris VA, Ojakangas CL, Fellows MR, Donoghue JP. Brain-machine and brain-computer interfaces. Stroke 2004; 35(11) (Suppl. 1): 2702-5.
[http://dx.doi.org/10.1161/01.STR.0000143235.93497.03] [PMID: 15486335]

[307] Lebedev MA, Nicolelis MA. Brain-machine interfaces: past, present and future. Trends Neurosci 2006; 29(9): 536-46.
[http://dx.doi.org/10.1016/j.tins.2006.07.004] [PMID: 16859758]

[308] Kurzweil R. The singularity is near: when humans transcend biology. New York: Viking 2005.

[309] Baker S. Rise of the cyborgs. Discover: Science, Technology, and the Future 2008; 52 http://discovermagazine.com/2008/oct/26-rise-of-the-cyborgs

[310] Clark A. Natural-born cyborgs: minds, technologies, and the future of human intelligence. Oxford: Oxford University 2003.

[311] Nicolelis MA. Brain-machine interfaces to restore motor function and probe neural circuits. Nat Rev Neurosci 2003; 4(5): 417-22.
[http://dx.doi.org/10.1038/nrn1105] [PMID: 12728268]

[312] Wolpaw JR, McFarland DJ. Control of a two-dimensional movement signal by a non-invasive brain-computer interface in humans. Proc Natl Acad Sci USA 2006; 101(17849): 54.

[313] Santhanam G, Ryu SI, Yu BM, Afshar A, Shenoy KV. A high-performance brain-computer interface. Nature 2006; 442(7099): 195-8.
[http://dx.doi.org/10.1038/nature04968] [PMID: 16838020]

[314] Hossain P, Seetho IW, Browning AC, Amoaku WM. Artificial means for restoring vision. BMJ 2005; 330(7481): 30-3.
[http://dx.doi.org/10.1136/bmj.330.7481.30] [PMID: 15626803]

[315] Kotchetkov IS, Hwang BY, Appelboom G, Kellner CP, Connolly ES Jr. Brain-computer interfaces: military, neurosurgical, and ethical perspective. Neurosurg Focus 2010; 28(5): E25.
[http://dx.doi.org/10.3171/2010.2.FOCUS1027] [PMID: 20568942]

[316] Schermer M. The mind and the machine: On the conceptual and moral implications of brain-machine

interaction. NanoEthics 2009; 3(3): 217-30.
[http://dx.doi.org/10.1007/s11569-009-0076-9] [PMID: 20234874]

[317] McGee EM, Maguire GQ. Implantable brain chips: ethical and policy issues. Med Ethics (Burlingt, Mass) 2001; 1-2: 1-2, 8.
[PMID: 15584168]

[318] Haraway D. A cyborg manifesto: science, technology, and socialist-feminism in the late twentieth century. Simians, cyborgs and women: the reinvention of nature. New York: Routledge 1999; pp. 149-81.

[319] Maheu MM. The future of cyber-sex and relationship fidelity: a brave new world booklet. Selfhelp Magazine 2006.

[320] Chatterjee A. The promise and predicament of cosmetic neurology. J Med Ethics 2006; 32(2): 110-3.
[http://dx.doi.org/10.1136/jme.2005.013599] [PMID: 16446417]

[321] Breitenstein C, Wailke S, Bushuven S, *et al.* D-amphetamine boosts language learning independent of its cardiovascular and motor arousing effects. Neuropsychopharmacology 2004; 29(9): 1704-14.
[http://dx.doi.org/10.1038/sj.npp.1300464] [PMID: 15114342]

[322] Stix G. Turbocharging the brain. Sci Am 2009; 301(4): 46-49, 52-55.
[http://dx.doi.org/10.1038/scientificamerican1009-46] [PMID: 19780452]

[323] Babcock Q, Byrne T. Student perceptions of methylphenidate abuse at a public liberal arts college. J Am Coll Health 2000; 49(3): 143-5.
[http://dx.doi.org/10.1080/07448480009596296] [PMID: 11125642]

[324] Baranski JV, Pigeau R, Dinich P, Jacobs I. Effects of modafinil on cognitive and meta-cognitive performance. Hum Psychopharmacol 2004; 19(5): 323-32.
[http://dx.doi.org/10.1002/hup.596] [PMID: 15252824]

[325] Randall DC, Viswanath A, Bharania P, *et al.* Does modafinil enhance cognitive performance in young volunteers who are not sleep-deprived? J Clin Psychopharmacol 2005; 25(2): 175-9.
[http://dx.doi.org/10.1097/01.jcp.0000155816.21467.25] [PMID: 15738750]

[326] Turner DC, Robbins TW, Clark L, Aron AR, Dowson J, Sahakian BJ. Cognitive enhancing effects of modafinil in healthy volunteers. Psychopharmacology (Berl) 2003; 165(3): 260-9.
[PMID: 12417966]

[327] BLTC Research. http://www.bltc.com/

[328] Giurgea C. Vers une pharmacologie de l'activité integrative du cerveau. Tentative du concept nootrope en psychopharmacologie. Actual Pharmacol (Paris) 1972; 25: 115-56.
[PMID: 4541214]

[329] Rose SP. Smart drugs: do they work? Are they ethical? Will they be legal? Nat Rev Neurosci 2002; 3(12): 975-9.
[http://dx.doi.org/10.1038/nrn984] [PMID: 12461554]

[330] Greely H, Sahakian B, Harris J, *et al.* Towards responsible use of cognitive-enhancing drugs by the healthy. Nature 2008; 456(7223): 702-5.
[http://dx.doi.org/10.1038/456702a] [PMID: 19060880]

[331] Volkow ND, Fowler JS, Logan J, *et al.* Effects of modafinil on dopamine and dopamine transporters in the male human brain: clinical implications. JAMA 2009; 301(11): 1148-54.
[http://dx.doi.org/10.1001/jama.2009.351] [PMID: 19293415]

[332] United Nations Drug Control Programme. Global illicit drug trends. New York : United Nations office for drug control and crime prevention Oxford 2001; pp. 1-282.

[333] Edelano L. Uber einige derivate den phenylmethocryhs ure wind die phenyhisobuttens ure. Berichte den Deutschen Chemischen Gesehlschaft 1887; 20(6): 16.

[334] Baberg HT, Nelesen RA, Dimsdale JE. Amphetamine use: return of an old scourge in a consultation psychiatry setting. Am J Psychiatry 1996; 153(6): 789-93.
[http://dx.doi.org/10.1176/ajp.153.6.789] [PMID: 8633691]

[335] Advokat C. Update on amphetamine neurotoxicity and its relevance to the treatment of ADHD. J Atten Disord 2007; 11(1): 8-16.
[http://dx.doi.org/10.1177/1087054706295605] [PMID: 17606768]

[336] Rose RM. Combating methamphetamine abuse. BJA act Sheet Office of Justice Programs Bureau of Justice Assistance 2009.

[337] Scott MS. Rave parties. Guide No14. US: Department of Justice. Office of Community Orientated Policing Services 2002; pp. 1-52.

[338] Reneman L, Booij J, Lavalaye J, *et al.* Use of amphetamine by recreational users of ecstasy (MDMA) is associated with reduced striatal dopamine transporter densities: a [123I]Iβ-CIT SPECT study-preliminary report. Psychopharmacology (Berl) 2002; 159(3): 335-40.
[http://dx.doi.org/10.1007/s00213-001-0930-0] [PMID: 11862367]

[339] Walker-Batson D, Smith P, Curtis S, Unwin H, Greenlee R. Amphetamine paired with physical therapy accelerates motor recovery after stroke. Further evidence. Stroke 1995; 26(12): 2254-9.
[http://dx.doi.org/10.1161/01.STR.26.12.2254] [PMID: 7491646]

[340] Grade C, Redford B, Chrostowski J, Toussaint L, Blackwell B. Methylphenidate in early poststroke recovery: a double-blind, placebo-controlled study. Arch Phys Med Rehabil 1998; 79(9): 1047-50.
[http://dx.doi.org/10.1016/S0003-9993(98)90169-1] [PMID: 9749682]

[341] Crisostomo EA, Duncan PW, Propst M, Dawson DV, Davis JN. Evidence that amphetamine with physical therapy promotes recovery of motor function in stroke patients. Ann Neurol 1988; 23(1): 94-7.
[http://dx.doi.org/10.1002/ana.410230117] [PMID: 3345072]

[342] Goldstein LB. Effects of amphetamines and small related molecules on recovery after stroke in animals and man. Neuropharmacology 2000; 39(5): 852-9.
[http://dx.doi.org/10.1016/S0028-3908(99)00249-X] [PMID: 10699450]

[343] Long D, Young J. Dexamphetamine treatment in stroke. QJM 2003; 96(9): 673-85.
[http://dx.doi.org/10.1093/qjmed/hcg113] [PMID: 12925723]

[344] Connemann BJ. Donepezil and flight simulator performance: effects on retention of complex skills. Neurology 2003; 61(5): 721.
[http://dx.doi.org/10.1212/WNL.61.5.721] [PMID: 12963782]

[345]  Holmes A, Heilig M, Rupniak NM, Steckler T, Griebel G. Neuropeptide systems as novel therapeutic targets for depression and anxiety disorders. Trends Pharmacol Sci 2003; 24(11): 580-8. [http://dx.doi.org/10.1016/j.tips.2003.09.011] [PMID: 14607081]

[346]  Kosfeld M, Heinrichs M, Zak PJ, Fischbacher U, Fehr E. Oxytocin increases trust in humans. Nature 2005; 435(7042): 673-6. [http://dx.doi.org/10.1038/nature03701] [PMID: 15931222]

[347]  Walker DL, Toufexis DJ, Davis M. Role of the bed nucleus of the stria terminalis versus the amygdala in fear, stress, and anxiety. Eur J Pharmacol 2003; 463(1-3): 199-216. [http://dx.doi.org/10.1016/S0014-2999(03)01282-2] [PMID: 12600711]

[348]  The National Institute of Mental Health. The numbers count: mental disorders in America No 01-4584. Washington, DC: NI 2003.

[349]  Kessler RC, Chiu WT, Demler O, Merikangas KR, Walters EE. Prevalence, severity, and comorbidity of 12-month DSM-IV disorders in the National Comorbidity Survey Replication. Arch Gen Psychiatry 2005; 62(6): 617-27. [http://dx.doi.org/10.1001/archpsyc.62.6.617] [PMID: 15939839]

[350]  Caldera EO. Cognitive enhancement and theories of justice: contemplating the malleability of nature and self. J Evol Technol 2008; 18(1): 116-23.

[351]  Moncrieff J. The myth of the chemical cure: A critique of psychiatric drug treatment. London: Palgrave 2015. [http://dx.doi.org/10.1057/9780230589445]

[352]  Edelman G. Second nature: brain science and human knowledge. New Haven, London: Yale University Press 2006.

[353]  Dinello D. Technophobia! science fiction visions of posthuman technology. Austin: University of Texas Press 2005.

[354]  Bostrom N. Are you living in a computer simulation? Philos Q 2003; 53(211): 243-55. [http://dx.doi.org/10.1111/1467-9213.00309]

[355]  Steels L. The Homo Cyber Sapiens, the robot Homonidus Intelligens, and the 'artificial life' approach to artificial intelligence. Burda Symposium on Brain-Computer Interfaces at Munich . 1995.

[356]  Johnston J. A future for autonomous agents: machinic merkwelten and artificial evolution Configurations 2002; 10(3): 473-516. [http://dx.doi.org/10.1353/con.2004.0004]

# SUBJECT INDEX

www.ingramcontent.com/pod-product-compliance
Lightning Source LLC
Chambersburg PA
CBHW041702210326
41598CB00007B/507